建筑立场系列丛书

司法建筑
Courthouses

[意]皮亚诺建筑事务所 等 | 编
唐瑞 | 译

大连理工大学出版社

004 基梅尔区_Schmidt Hammer Lassen Architects

008 新巴塞罗那市档案馆_OP Team + Mendoza Partida + Ramon Valls Architects

012 昂热圣塞尔吉斯码头“变形记”_Hamonic + Masson & Associés

016 统营“Mare Camp”_HENN + Posco A&C

Cultural Warehouses文化仓库

020 文化仓库和新工业范例_Davide Pisu

026 罗特曼谷物升降机_KOKO Architects

036 创新动力工场_Atelier van Berlo + Eugelink Architectuur + De Bever Architecten

048 北京文化创新工场_Cobblestone

066 阿尔斯通仓库修复项目_Franklin Azzi Architecture

082 维堡街头圣地_EFFEKT

Courthouses司法建筑

100 理想的实现和正义的紧张_Herbert Wright

108 拉里奥加法院_Pesquera Ulargui Arquitectos

126 利摩日法院_ANMA

138 科尔多瓦司法大楼_Mecanoo Architecten + AYESA

154 巴黎法院_Renzo Piano Building Workshop

172 富瓦法院_Philippe Gazeau

190 自由空间，第16届威尼斯建筑双年展_Tom van Malderen

198 建筑师索引

004 Kimmel Quarter_Schmidt Hammer Lassen Architects

008 New City of Barcelona Archives_OP Team + Mendoza Partida + Ramon Valls Architects

012 "Métamorphose" in Quai Saint-Serge, Angers_Hamonic + Masson & Associés

016 Tongyeong "Camp Mare"_HENN + Posco A&C

Cultural Warehouses

020 Cultural Warehouses and the New Industrial Paradigm_Davide Pisu

026 Rotermann Grain Elevator_KOKO Architects

036 Innovation Powerhouse_Atelier van Berlo + Eugelink Architectuur + De Bever Architecten

048 Beijing Cultural Innovation Park_Cobblestone

066 Alstom Warehouse Renovation_Franklin Azzi Architecture

082 Streetmekka in Viborg_EFFEKT

Courthouses

100 The Materialisation of the Ideals and Tensions of Justice_Herbert Wright

108 La Rioja Courthouse_Pesquera Ulargui Arquitectos

126 Limoges Courthouse_ANMA

138 Palace of Justice in Córdoba_Mecanoo Architecten + AYESA

154 Paris Courthouse_Renzo Piano Building Workshop

172 Palace of Justice in Foix_Philippe Gazeau

190 Freespace, the 16th Venice Architecture Biennial_Tom van Malderen

198 Index

基梅尔区

Kimmel Quarter _ Schmidt Hammer Lassen Architects

Schmidt Hammer Lassen (SHL建筑师事务所) 在基梅尔区的建筑设计竞赛中拔得头筹。基梅尔区占地11 500m², 几乎占据了里加中区的整个城区。基梅尔区原先容纳了基梅尔啤酒厂, 这是一座19世纪的啤酒酿造厂, 现已大面积荒废停用, 但却有着深厚的历史底蕴。该项目的设计愿景就是要将此地打造成具有深厚历史底蕴、充满活力的城区。

获胜方案的设计概念集中体现在一座建筑面积达30 000m²的办公楼上, 该办公楼是一座开放式的建筑, 能够直接连通到具有城市气息的室内外公共区域, 其设计很吸引人。该方案将尽可能保持现有建筑物不变, 以凸显里加地区真正的历史底蕴与特点。新办公楼的地面层建筑灵感来自原啤酒厂的拱门, 由改建现场回收的砖块建造而成, 其材料极具历史感。

从新办公楼的各个楼层都可以进入露台和屋顶花园等户外空间, 因此该建筑的使用者将有更多机会亲近自然。此外, 新设计将原基梅尔啤酒厂的工业后院改造成了一个充满活力的广场, 新墙面使用了更多的回收砖块, 再加上郁郁葱葱的园林景观、砖凳木椅以及从屋顶取水的镜像水元素, 使建筑宛若重获新生。通过建造美丽、引人入胜的庭院和广场, 连接新旧元素, 该建筑的整体设计将打造出有着经久不衰的经典外观以及独特的当代风韵, 并且极具都市感的建筑风格。

作为一座充满雄心壮志的城市, 里加制定了长期和短期的战略, 例如, 完成欧盟2020年气候和能源一揽子目标, 成为一座可持续发展城市。而基梅尔区矩形网格外立面设计允许了最大量的阳光进入建筑内部, 与之相结合的薄层又使建筑避免了阳光直射, 该项目因此成为里加市未来城市发展的典范。

Schmidt Hammer Lassen Architects won a competition to design Kimmel Quarter, an 11,500m² area that assumes nearly an entire city block in Riga's Central District. Kimmel Quarter is the former home of Brewery Kimmel, a 19th century beer brewery mostly abandoned yet rich in history. The project called for a design that realizes the vision of a new vibrant city block with deep historical roots.

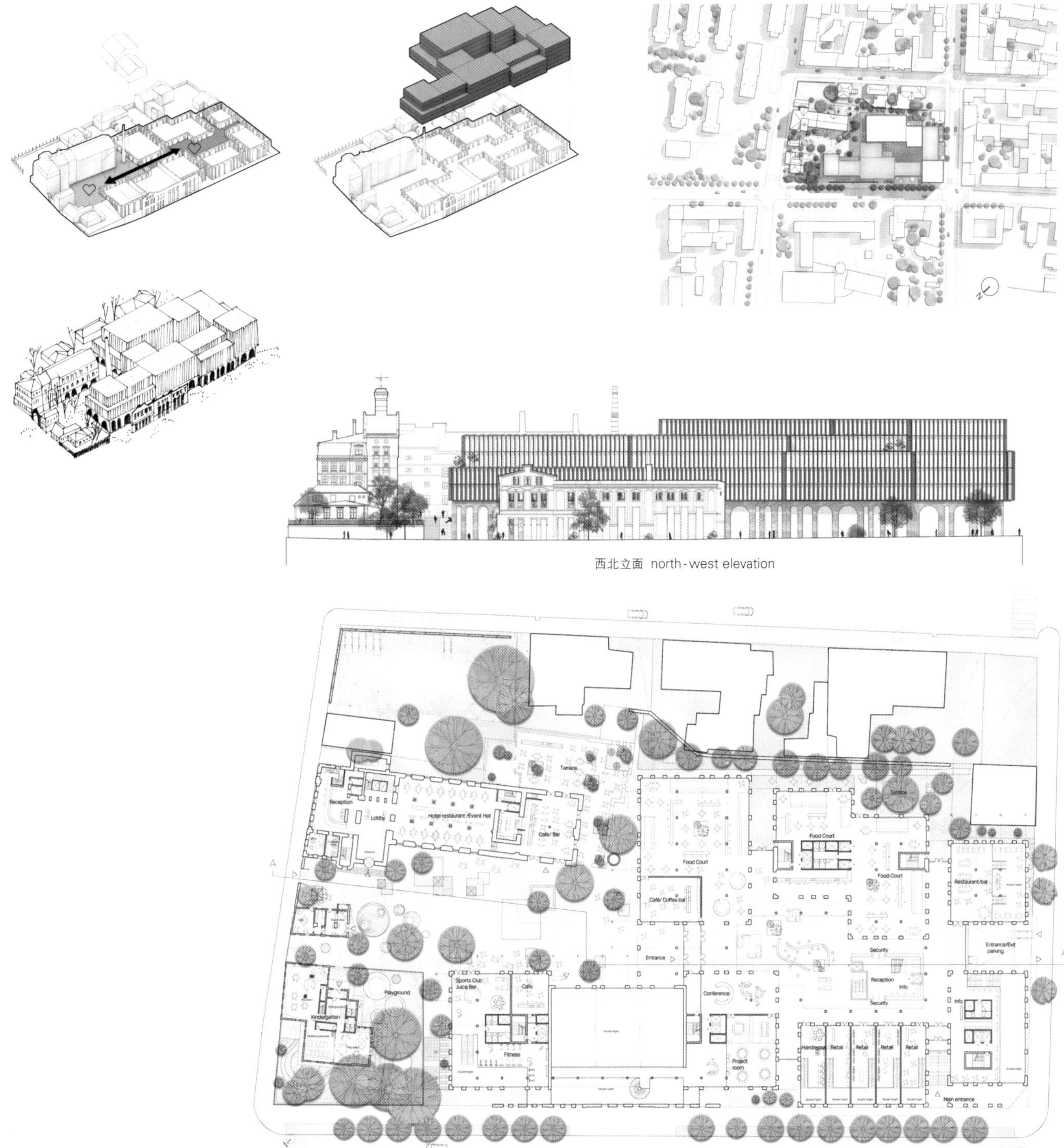

西北立面 north-west elevation

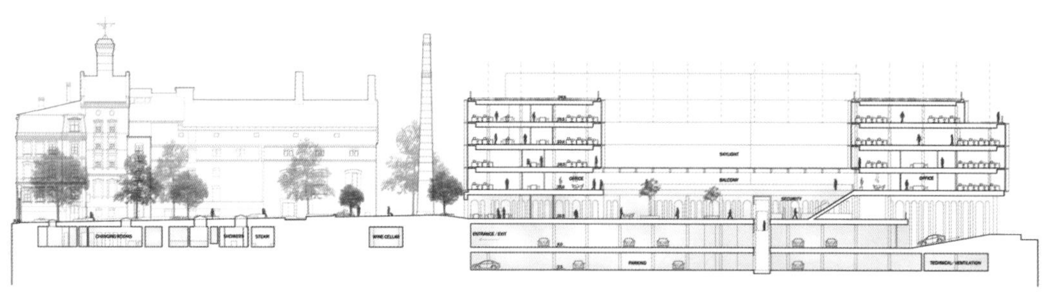

A–A' 剖面图 section A-A'

The concept of the winning proposal centred on a 30,000 m² office building that is open and inviting with direct access to indoor and outdoor public areas with urban qualities and atmosphere. The existing buildings will remain as unaltered as possible, enhancing the authentic character of Riga's historical fabric. The new office building's ground floor is inspired by the arches of the brewery, and with the use of recycled bricks from the site, its material traces back to history.

Outdoor spaces such as terraces and roof gardens allow access to the outdoors from all levels and bring users of the building closer to nature. In addition, the new design turns the Brewery Kimmel's former industrial backyard into a vibrant plaza revitalized with a new surface made of more recycled bricks, lush landscaping, brick and timber benches, and mirroring water elements fed with water from the roofs. By creating beautiful, inviting courtyards and plazas that connect the old and the new, the overall design will create a strong urban expression with a timeless, classic appearance that is also uniquely contemporary.

Riga is a city with ambitions of transitioning into a sustainable city with long and short-term strategies like reaching the European Union's 2020 climate & energy package goals. Combining rectangular grid facades that allow maximum daylight deep into the building and lamellas that provide shading from direct sunlight, Kimmel Quarter will become an example for the future development of the city of Riga.

巴塞罗那新城市档案馆

New City of Barcelona Archives_OP Team + Mendoza Partida + Ramon Valls Architects

OP Team、Mendoza Partida和Ramón Valls Architects赢得了由BIMSA组织的两阶段的建筑设计竞赛，因此得以负责巴塞罗那市档案馆所在地的Can Batlló旧址8号仓库的修复工程。除了保护并且升级建筑的特色元素之外，该项目还旨在将档案馆改造成一座能够容纳50km文档、400万张照片以及数千图表和图片文件的新建筑，而这些记录着巴塞罗那历史的资料现存于21个不同的档案存储中心。

这个名为"集会（Agora）"的获奖项目展示了一座与周围环境紧密融合、光线充足的建筑。它虽然外观简单，却有一个结构复杂的档案室，并采取了以人为本的设计原则。该项目主要设计建造了位于仓库中央区域北部的储存库，那是一个宽27m、长135m、层高6m的储存空间，由两个体量构成，共有三层，同时建筑师为了扩大公共区域，又设计了仓库南半部以及围绕它的建筑环形结构，这个环形结构宽14m，高一层。同时，建筑师又建造设计了三个内部庭院，进一步扩大了公共空间。这个空间为建筑提供了日光和自然通风，使人们可以看到有趣的景色，并显示出建筑物大大小小的原有结构。从三个庭院中最大的那个庭院走过就能到达档案馆的主入口，这个庭院也是通往中庭及建筑地面层的公共广场"集会"的门户。经由该广场可以通往建筑综合体在中央大楼周围的各种社交和文化空间（画廊、礼堂、会议室）。而一层和二层则用作咨询服务和内部工作空间。集会广场四通八达，多条路线皆可到达，该广场将Gran Via与La Bordeta社区之间的道路，与新设施和Can Batlló区新公园之间的道路连接了起来。

建筑师修复了建筑的外立面和内部空间，并且复原了屋顶的斜坡，突出了原有建筑的体积和宽敞的空间。档案馆存储库建有48个独立小房间，以便于根据每种类型的文件所需的环境条件进行单独存放。该建筑又通过建立缓冲空间与现有建筑隔离开来，以尽量减少它对外部能源的需求。为了实现自给自足，该项目采用了地热和太阳能光伏能源等可再生能源。

OP Team, Mendoza Partida, and Ramón Valls Architects won the two-phase competition organised by BIMSA for the rehabilitation of warehouse eight on the Can Batlló site as home to the City of Barcelona Archives. In addition to conserving and upgrading the building's characteristic elements, the aim of the competition was to adapt it to house the over 50km documentation, four million photographs, and thousands of graphics and cartographic documents that illustrate and document the history of Barcelona, currently kept in 21 different centers.

Entitled "Agora", the winning project presents a light-filled building that is rooted in its surroundings, apparently simple despite the organisational complexity of an archive,

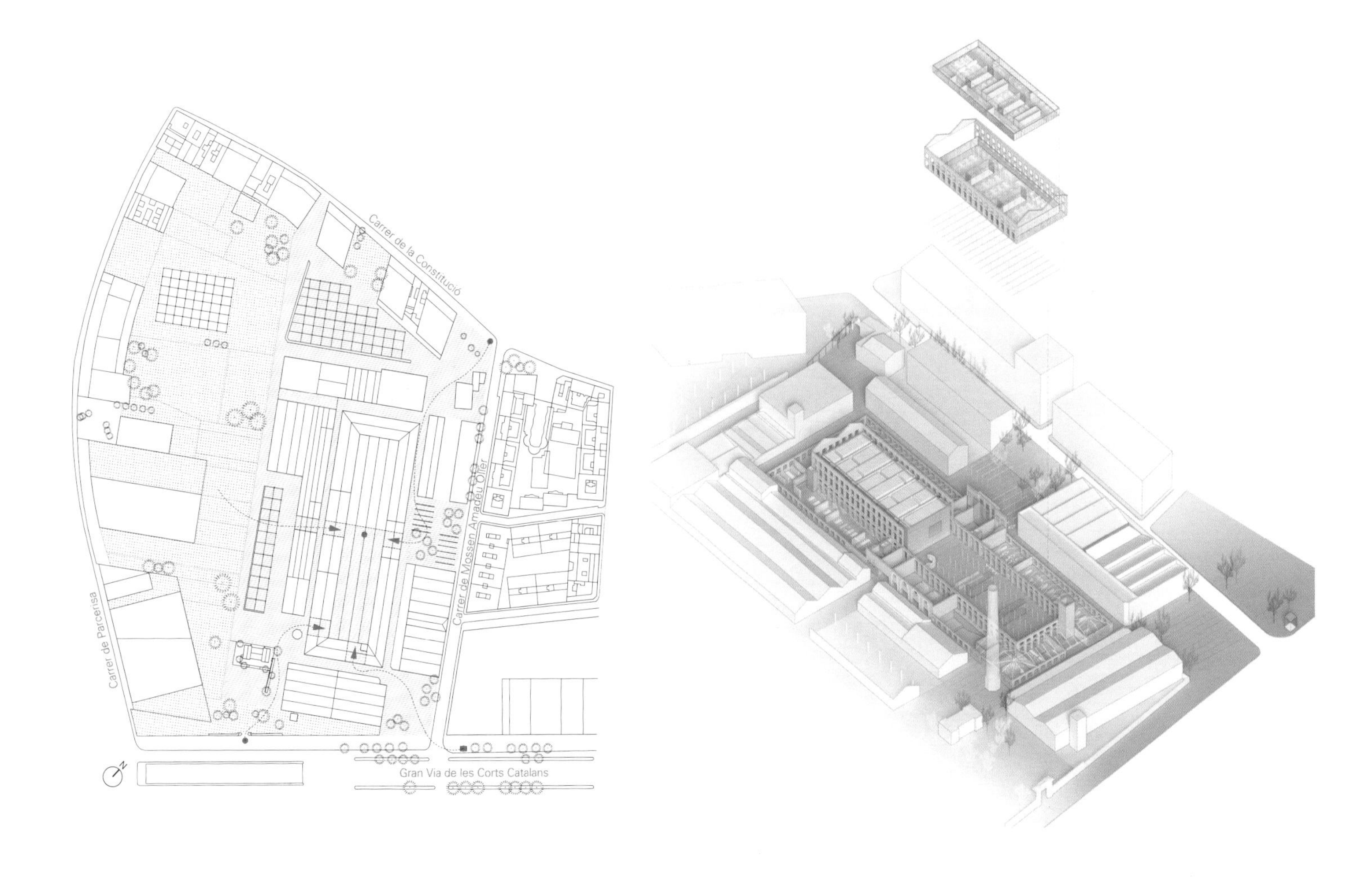

东北立面 north-east elevation

a building that places people at the center of its design. It concentrates the area containing the repositories in the northern half of the central part of the warehouse, a two-volume space, 27m wide, 135m long with three six-m-high levels, and designs the southern half and the built ring around it, 14m wide with one floor, for more public areas. This space for the public is further enhanced by three inner courtyards that provide daylight and natural ventilation, offer interesting lines of sight, and reveal the original structure of the building in all its magnitude. The main entrance to the archives is via the largest courtyard, which acts as an atrium and threshold to the public square, the Agora, that occupies the ground floor. This square provides access to the complex's various social and cultural spaces surrounding the central building (galleries, auditorium, seminar rooms). First and second floors are given for consultations and in-house work. The Agora is permeable, allowing multiple alternative approaches, connecting the routes that link Gran Via with the neighbourhood of La Bordeta, and the new amenities with the new park at Can Batlló.

The project recovers the facades and inside space, and restores the slope of the roof, highlighting the volume and the sheer spaciousness of the original building. The archive repositories are housed in a structure built by adding 48 independent cells to facilitate the environmental conditions required by each type of document. This structure is, in turn, separated from the existing construction by buffer spaces to minimise the need for external energy. With the aim of self-sufficiency, the project incorporates renewable options such as geothermic and solar photovoltaic energies.

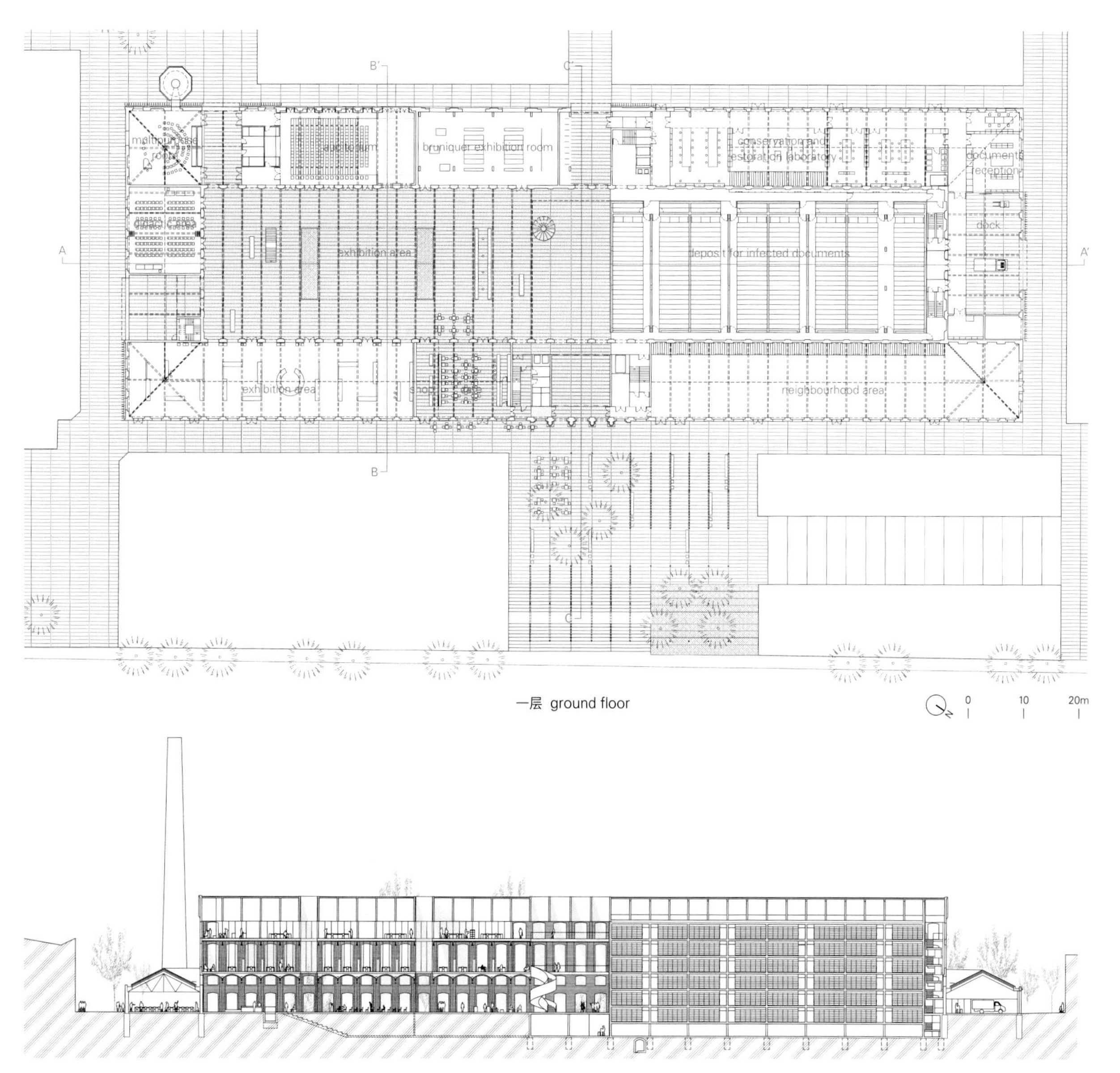

一层 ground floor

A–A' 剖面图 section A-A'

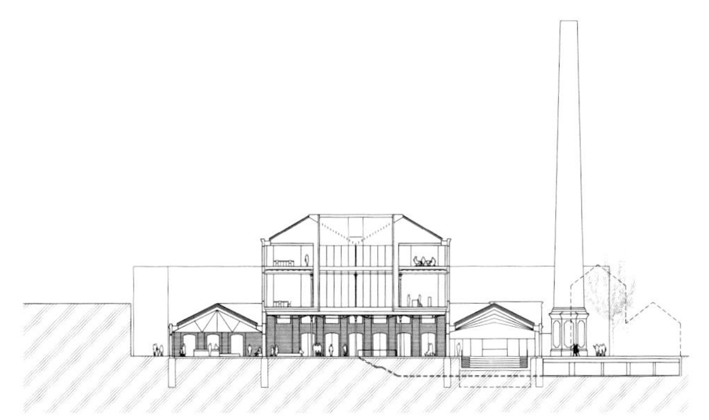

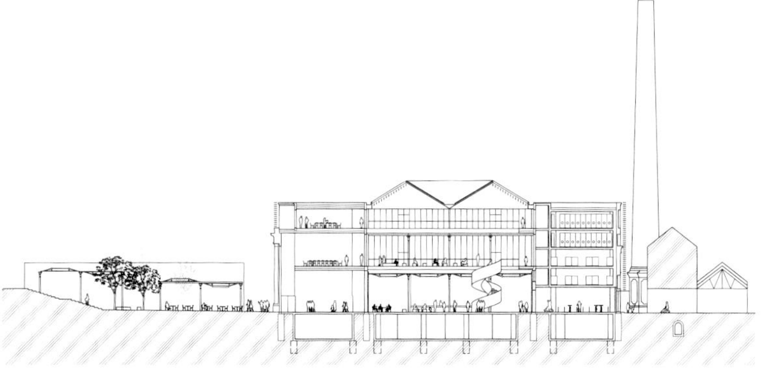

B-B' 剖面图 section B-B'

C-C' 剖面图 section C-C'

昂热圣塞尔日码头“变形记”

“Métamorphose” in Quai Saint-Serge, Angers_Hamonic+Masson & Associés

名为“昂热设想”的国际建筑设计竞赛要求参赛团队选择法国昂热地区的六个场地之一，创作一套具有创造性的建筑设计方案。该项目要求将其中最大的一片场地，即一片工业棕地，改造成一个充满活力和创新的社区，致力于高等教育和城市活动。由Hamonic+Masson& Associés建筑师事务所设计的“圣塞尔日码头”获选为本次竞赛的获胜方案。

在该方案的设计之初，建筑师就对这一场地的独特性进行了思考，并意图建立一个象征城市雄心壮志的项目，为法国及其他欧洲大城市注入活力。该项目将创新作为概念工具，来改造这片未知的地域。在城市景观方面，创新就是要找到与公共空间和谐相处的新方式，创新也是一种吸引人的手段。建筑师试图寻求新的方案以及技术和空间解决办法，打造出一座强有力的标志性建筑。独特的选址及其周边环境，例如，共享居住、攀岩墙、体育中心和健身房、办公室、餐厅和停车场等，也为建筑增光添彩。

“圣塞尔日码头”是一个建筑综合体，与其所在的场地融为一体，并期望成为一种创造冒险、惊喜和情感的新型建筑。与此同时，建筑师也渴望将该项目打造成一个代表城市雄心的建筑。该建筑在设计中引入了“流动性”，将填充空间和空白空间视为不同的地形层次进行连接，从而避免了断裂的情况。该建筑因此成为该领域的突出性建筑。

因此，这是一个融合的项目。建筑师不希望特立独行或是挑战传统，而是想通过创造对话并重建连接来进行创造。这样的对话和交流将打造出一个开创性项目，一个21世纪的象征：“变形记”项目。

International architecture competition “Imagine Angers” asked design teams to propose an innovative architecture on one of six different sites in Angers, France. The brief called for a reconversion of this industrial brownfield zone, one of the biggest sites, into a dynamic and innovative neighbourhood, dedicated to higher education and urban activity. “Quai Saint-Serge” by Hamonic + Masson & Associés was selected as the winning proposal for the competition. The proposal began with the reflection on the exceptional

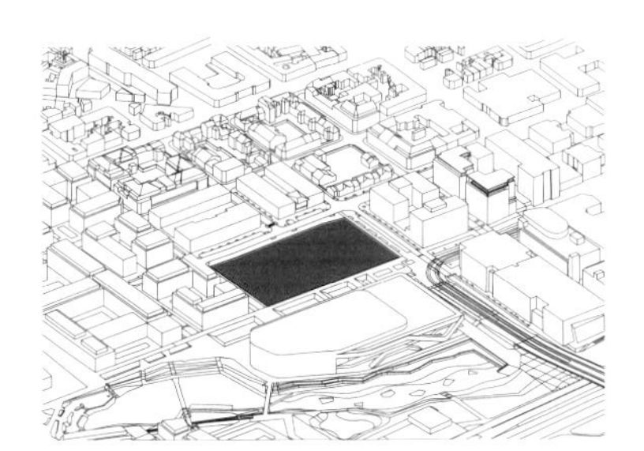

有象征意义的场地
an emblematic site

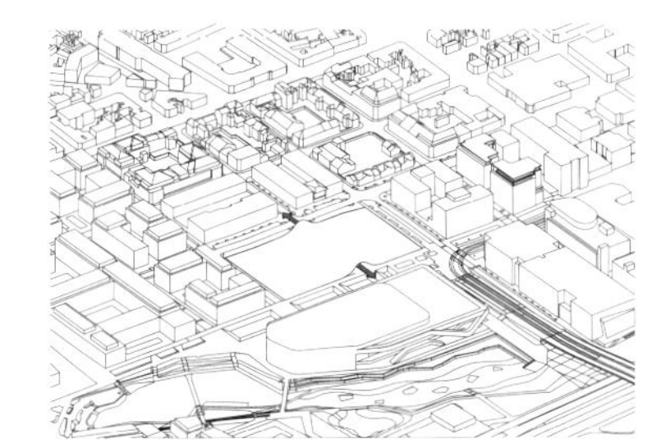

通道
a passage

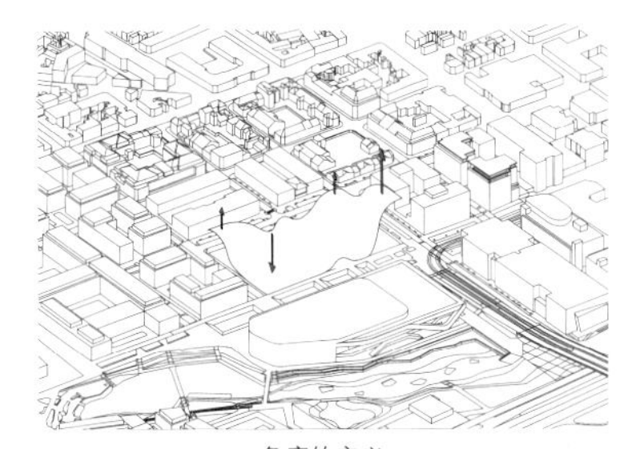

角度的定义
definition of angles

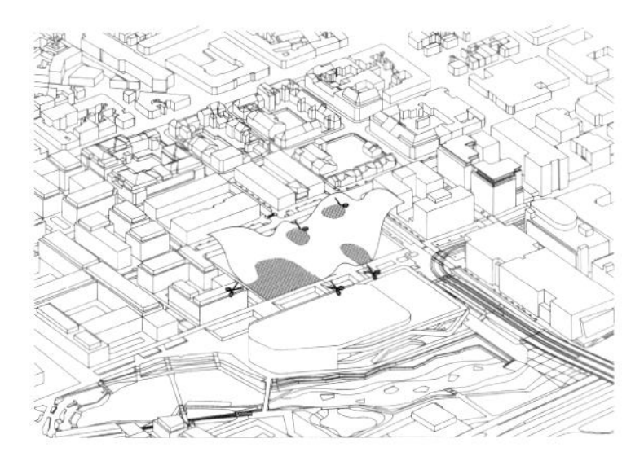

在体量中创造缝隙
create gaps in the volume

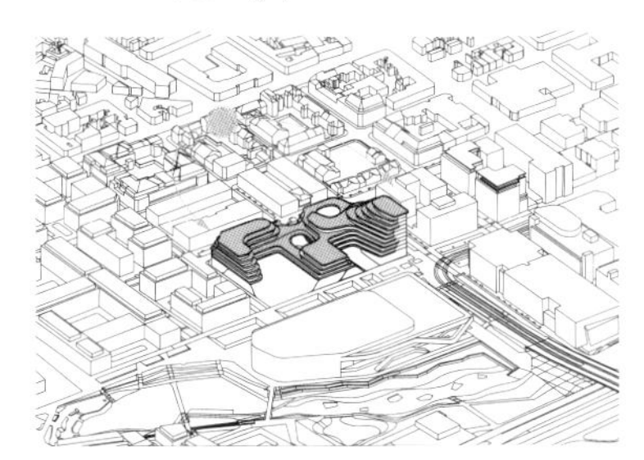

根据定向调整形态
adapt morphology according to orientation

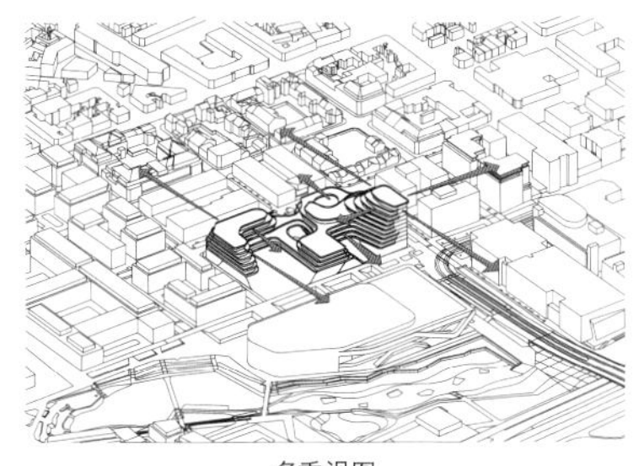

多重视图
multiple views

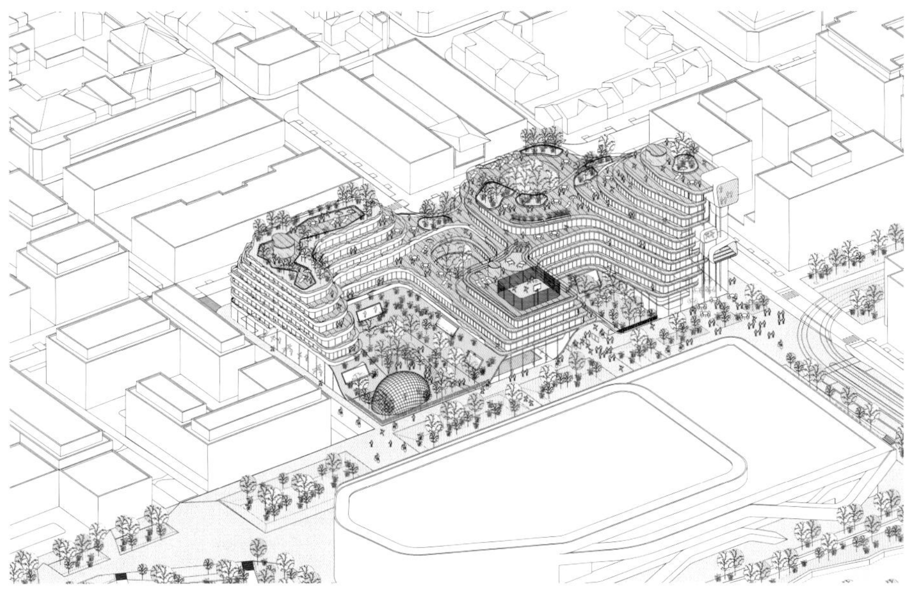

1. 热带温室
2. 工作坊
3. 接待处
4. 巴黎地方法院
5. 人造实验室
6. 当地自行车
7. 大厅
8. 当地商业区
9. 休息区
10. 酒吧
11. 厨房
12. 更衣室
13. 健身室
14. 健身接待处
15. 游泳池
16. 水上活动区
17. 啤酒厂
18. 联合办公区/会议室
19. 更衣室
20. 攀岩室
21. 办公室接待处
22. 餐厅

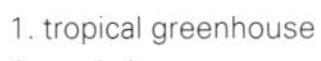

1. tropical greenhouse
2. workshop
3. reception
4. Conciergerie
5. fab lab
6. local bike
7. grand hall
8. local commercial
9. break area
10. bar
11. kitchen
12. cloakrooms
13. fitness
14. Physic form reception
15. swimming pool
16. aquatic area
17. brewery
18. co-working/meeting
19. changing rooms
20. climbing room
21. office reception
22. restaurant

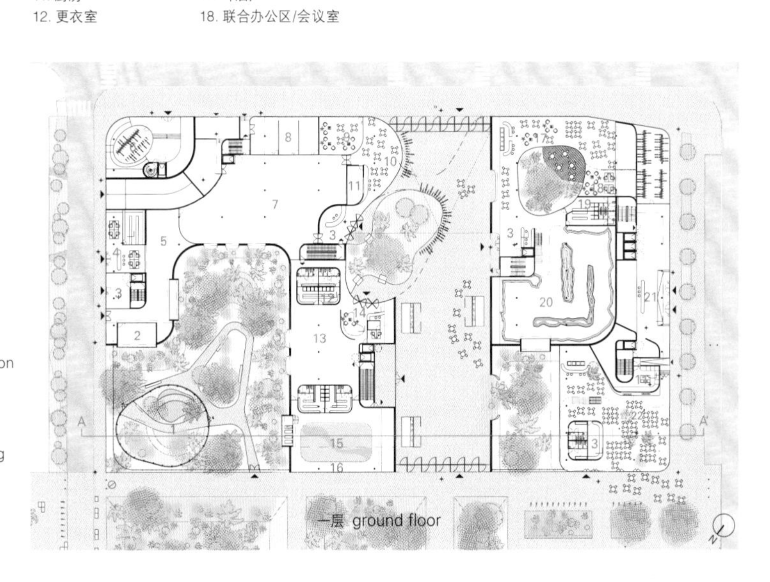

site, aiming to build a project that will be emblematic of the city's ambition, animating all national and European cities. Innovation was employed as a conception tool to transform unknown places and territories. When it comes to urban landscape, innovation is about finding new ways of living in harmony with the public space, as well as a tool for attractivity. This research provided the urge to look for new programmes, technical and spatial solutions, resulting in the project to become a strong signal. The context combined with this exceptional site enhanced the project with programmes like co-living, climbing wall, sports center and gym, offices, restaurant, and parking.

"Quai Saint-Serge" is a synthesis that will implant itself in the site and become a new typology that creates adventure, surprise, and emotions alongside the willingness to conceive a project that represents the city's ambitions. "Fluidity" is introduced in the design by linking filled and empty spaces and treating them as different topographical levels, avoiding ruptures. The building thus becomes an extruded piece of the territory.

It is therefore a confluence project. The architects did not wish to differentiate or to provoke, but to create a dialogue and to reconnect in order to invent. This exchange will create a pioneering project, a symbol of the 21st century: the project Métamorphose.

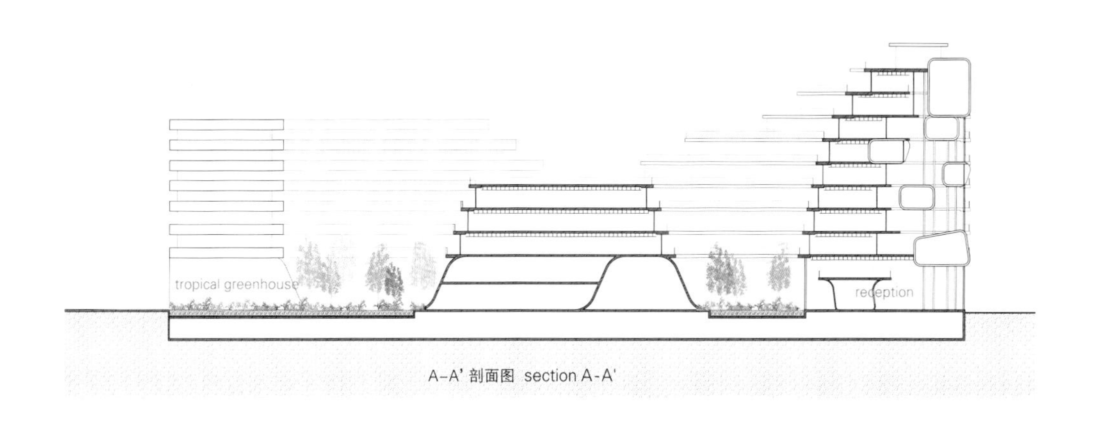

A–A' 剖面图 section A–A'

统营“Mare Camp”

Tongyeong “Camp Mare” _HENN + Posco A&C

在韩国统营船坞城市再生项目Camp Mare建筑设计竞赛中，德国海茵建筑设计公司和韩国浦项钢铁公司A&C建筑设计公司的联合设计方案获得了一等奖。这是一项由韩国LH公司支持并推动的国际竞赛，作为韩国通过量身定制的振兴战略来努力发展沿海重点城市的一部分。该获奖设计方案是由德国海茵建筑设计公司开发的，并与韩国浦项钢铁公司A&C建筑设计公司、Metaa、Inwooplan景观设计事务所、Siteplanning建筑事务所、SLA工程建筑公司、德勤安永和Yooshin工程公司合作完成的。

该项目位于韩国的统营市新区Camp Mare。该场地位于原先造船厂的旧址上，设计团队对这些工业遗产进行了再次利用，同时加入一系列新的建筑和设施，打造出了全新的集手工艺、旅游、研发和生活于一体的城市中心。

Camp Mare场地北边邻水，而南边毗邻风景如画的米鲁克山。设计团队构想了两个直线型的海滨扩建地带，并涵盖曾经的造船厂地区，将南部的山地景观通过两个公园与海滨区域相连。整个项目分为五个不同的区域，每个区域都有自己的独特之处，由不同的设施和建筑构成，它们将整个热闹的公共海滨地带连接起来。

统营市一直作为一个制造型城市存在，这也成了Camp Mare项目

设计的出发点。该项目融入了城市的传统工艺、烹饪文化和造船传统，并通过引进新技术、研究设施、文化活动、旅游景点等，重新激活这一片的海滨生活。该规划还包括"12所学校"项目，这些学校将成为推动创新的标志，为整个城市和地区创造一个全新的工业生态系统。每个区域内都包含多所学校，每所学校的教育都将关注工业循环的各个阶段：研发、设计、生产、测试、营销和销售。

Camp Mare最终将成为统营市一块五彩缤纷、充满活力的新区，将为统营市的再生与未来发展奠定基础。

HENN + Posco A&C won the 1st Prize for the Urban Regeneration of Tongyeong Dockyard in South Korea – Camp Mare. It was an international competition promoted by the Korean LH Corporation as part of South Korea's national effort to develop key coastal cities by means of tailored revitalisation strategies. The winning proposal was developed by HENN, the design lead of the consortium with Posco A&C, Metaa, Inwooplan Landscape Architects, Siteplanning

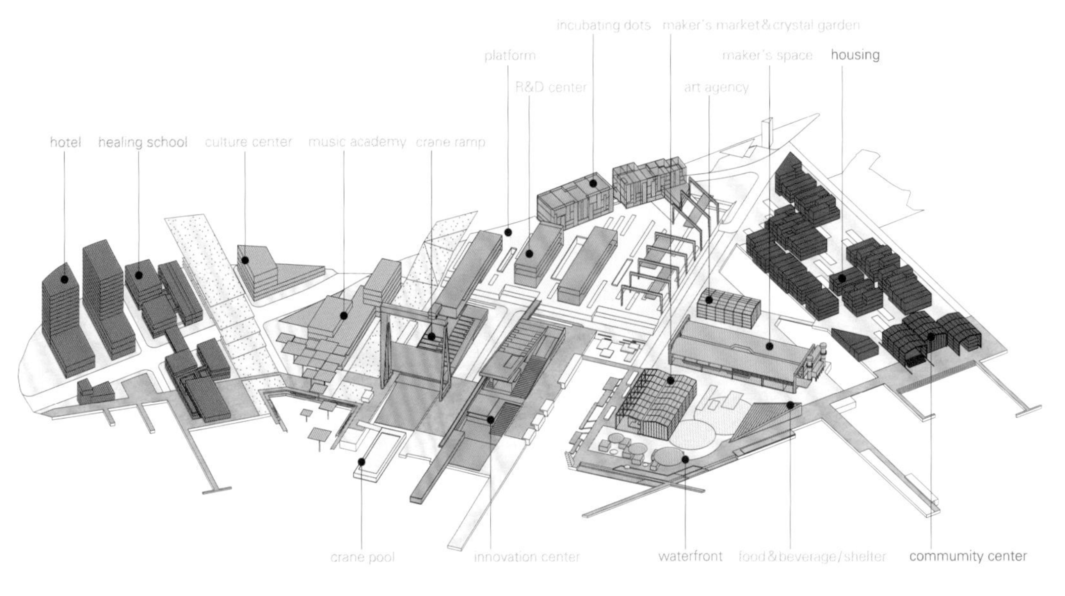

Architects, SLA Engineering & Architecture, Deloitte Anjin, and Yooshin Engineering Corporation.

Camp Mare is a new district in the city of Tongyeong. Located on the site of a former shipyard, it celebrates its heritage by reusing several of the industrial structures as well as introducing a range of new programmes and buildings to create a hub for craftsmanship, tourism, research and development, and living.

Camp Mare's site is bordered to the north by the water and to the south by the picturesque Mireuk Mountains. The waterfront is extended by the addition of two rectilinear bays that carve into the dockyard. The mountainous landscape in the south is connected by two parks to the waterfront. The project is divided into 5 distinct districts with a mixture of programmes and buildings, all connected by a lively public waterfront.

Tongyeong's identity as a maker's city is the DNA of Camp Mare. It borrows the city's traditional arts and crafts, its culinary culture and its tradition in shipbuilding. This rich heritage is complemented by the introduction of new technologies, research facilities, cultural activities, touristic attractions and waterfront living.

The overall program consists of the so-called "12 Schools". Their goal is to drive innovation and to generate a new industrial ecosystem for the city and the region. Each neighbourhood houses multiple schools that revolves around an industrial cycle: research and development, design, production, testing, marketing, and sales.

Camp Mare will become a dynamic, colourful, and lively new district that sets the basis for the regeneration and future development of Tongyeong.

TONGYEONG
MAKER'S SPACE
통영
메이커 스페이스
크리스탈가든

SHINAsb
통영시
LIVING MUSEUM

文化仓库

Cultural W

本文描述了如何运用不同的设计策略，将已停用的工业建筑改造成具有其他功能性的建筑。为此，本文分析了五个当代后工业转型项目，并从三个方面阐述了不同的建筑设计理念和态度。第一个方面是旧工业建筑与其"改造再利用"所需新元素之间的空间关系，二者相互包裹或是相互交叉。第二个方面是现有建筑和新建筑不同设计理念的相互作用：例如，相对于那些稀奇古怪的高科技产品来说，什么才是真正的纪念碑呢？第三个方面描述了新旧之间不可避免的摩擦是如何表现、隐藏或被忽视的。本篇文章将通过五个建筑设计项目为读者展现一系

This article depicts how different design strategies can be implemented in the conversion of an industrial building to other functions after it has ceased its activities. To this end, five contemporary post-industrial conversion projects are analysed, and different stances are described in relation to three aspects. The first is the spatial relationship between the old industrial buildings and the new elements needed for their "adaptive reuse", whether one is encased into the other or they are mutually intersected. The second is the interaction between the different codes of existing and new: for instance, what is often a stark monumentality vis-à-vis to new outlandish hi-tech additions. The third aspect describes how the inevitable frictions between the old and the new are exhibited,

罗特曼谷物升降机_Rotermann Grain Elevator/KOKO Architects
创新动力工厂_Innovation Powerhouse/Atelier van Berlo + Eugelink Architectuur + De Bever Architecten
北京文化创新工场_Beijing Cultural Innovation Park/Cobblestone
阿尔斯通仓库改造_Alstom Warehouses Renovation/Franklin Azzi Architecture
维堡街头圣地_Streetmekka in Viborg/EFFEKT
文化仓库和新工业范式_Cultural Warehouses and the New Industrial Paradigm/Davide Pisu

arehouses

列不同的设计方法，而这些设计体现出了建筑师对城市工业综合体的生命周期及其转型的更加微妙的理解。也就是说，这些设计方法来源于生产场所或者有形物质的转化场所，因此，它们围绕着一系列有逻辑的、重复发生在一个严格空间维度上的活动进行组织。而这些设计方法同时也来源于文化消费、社会互动和商业培育的场所，而与之相反，这些场所受到了随意行为和不稳定构造的影响。简而言之，这些新的建筑设计方法代表了从物质到非物质的经济转型，从结构性的、坚固的硬结构到轻便、流动、不稳定的软结构的转型。

hidden or neglected. The five projects presented here offer a set of different approaches that relate to a subtler understanding of the life cycle of urban industrial complexes and their transition. These namely stem from places for the production or transformation of material goods; hence, they are organized around a logical series of repetitive activities that takes place in a rigid spatial dimension, to places of cultural consumption, social interaction and business fostering which are, in contrast, subject to extemporary actions and unstable configurations. In short, they represent the economic transition from the material to the immaterial, from the hardware – structural, strong organization – to the software – light, liquid and unstable.

文化仓库和新工业范式

Cultural Warehouses and the New Industrial Paradigm

Davide Pisu

法国哲学家阿兰·图海纳于1968年提出了“后工业社会”的概念，当时富裕的西方社会正在经历影响深远的转变：换句话说，第三产业不断扩张，著名的学生革命浪潮也一触即发。这两个明显不相关的事件却传达出了同样的信息，那就是重工业时代即将结束。而图海纳也预见到了这一点：一方面，新型的、非物质形态的生产方式正在不断取代传统的手工劳作；另一方面，新一代人开始拒绝劳动分工，转而青睐其他经济模式，尤其是文化生产。居伊·德波的《景观社会》和鲁尔·瓦纳格姆的《日常生活的革命》都为我们揭示了这样的现实，他们试图通过建筑设计，将个体的文化体验从“异化的具体制造”中拯救出来。[1]在当前的情况下，尽管自动化是大势所趋，重工业和旧式手工劳作仍然是全球经济的基本组成部分，并且逐渐瓦解了一度在工业界蓬勃发展的文化。

现如今，人们已经将重工业装配生产线视为旧物，而重工业建筑也见证了人们脱贫致富的光荣时刻，并成了那个时代的遗产。因此改造废弃工业旧地也成为建筑设计中一个常见的主题。从都灵的灵格托工厂到诺曼·福斯特的红点设计博物馆，不论他们是试图保存，还是揭示旧建筑的工业过往，不论是想改造工业旧建筑的内部，还是想建造一个崭新的外壳保护内部，不论是想颠覆旧建筑的布局和个人体验，还是仅仅想改造旧建筑使其与新的生活方式相适应，建筑师在设计中感怀过去并不是因为这些建筑本身具有历史价值，而是因为旧工业在世界历史中曾经具有的重要意义。

由考斯顿公司设计并于2015年建成的北京文化创新工场就是这样的建筑范例（48页）。该项目改建于20世纪90年代初一座较新的锅炉煤炭热力厂。建筑设计师团队对原有建筑进行了颠覆性改造，保留并使其内部结构可见，作为对其珍贵的工业历史的见证。在这个过程中，建筑师将该改建场地视为一个重写本，通过新建筑，一系列发生在不同历史时期的故事也得以在同一实体空间中集中展现。建筑师将原有建筑的外墙彻底推翻重建，使用了技术水平中等的灰色幕墙，建筑内部结构清晰可见。建筑外立面不同形状的结构具有不同的功能，一块悬挑玻璃面向周围的景观，这些现代化的建筑设计掩盖掉了建筑的历史过往。

When in 1968 the French philosopher Alain Touraine wrote about the "Post-industrial society", the wealthy Western world was undergoing a deep transition; namely, a shift toward the expansion of tertiary, and the famous student revolutionary season was on the verge of its beginning. These two apparently unconnected events carried the signs of the end of heavy industry, which Touraine foresaw: from one side, new and less material forms of production were increasingly taking over traditional manual labour, whilst on the other side, the new generations started refusing the division of labour in favour of other economic paradigms, and in particular the cultural production. It was a story told – amongst others – through the revealing works of Guy Debord's *The society of the spectacle* and Raoul Vaneigem's *Traité de savoir-vivre à l'usage des jeunes générations*, which sought to save the individual cultural experience from the "concrete manufacture of alienation".[1] In this scenario, heavy industry and old manual labour were still a fundamental part of the global economy, though increasingly shifting toward automation, when not offshoring, and gradually dissolving the culture that once thrived in the industrial world.

Today, the heavy industry assembly line is considered something relative to our past, and its buildings are seen as a legacy of a moment of glorious emancipation from poverty. Architectural projects for disused or neglected industrial sites are a common architectural theme. From Turin's Lingotto to Norman Foster's Red Dot Design Museum, whether they try to preserve or expose the industrial past, operate on the interior of the old envelope or protect it with a brand new gleaming shell, or radically change the distribution and the individual experience of the old or simply adapt it to a new life, they celebrate the old past not for the historic value of the building themselves but for the significance the old industry once had in the world's history.

One such building is Cobblestone's Beijing Cultural Innovation Park (p.48), which was realised in 2015. The renovation was operated on a relatively new coal heating station, built just at the beginning of the 1990s. The architects created a drastic transformation of the existing building, deciding to save and keep visible the inner part of the complex that testified to its previous industrial life. In doing so, the architects considered the site as a palimpsest, whereby the architecture becomes the place where a number of layered stories are combined in the same physical

因此，北京文化创新工场向我们展示了两个不同的方面。从街面上看去，反光的建筑表皮传递出明亮性、金属感以及温和的创新。而从建筑内部及庭院一侧看去，整座建筑的特色是采用了旧式的工业红砖墙，加上红色的考顿钢屋顶扩建结构。屋顶采用了棱角分明的几何图形来宣告改造理念，以凸显旧式沉重和新式轻便结构之间的反差。原建筑被保留下来的部分运用了特殊的色彩和形状，除此之外，整座建筑中似乎并无新旧之间的不和谐之处。建筑师精心设计了每个细节，同时每个疏漏的细节（例如，铁锈、缺失的边角）也都被修复了。

KOKO建筑师事务所设计的罗特曼谷物升降机（26页）位于爱沙尼亚塔林地区的历史中心。原始建筑是20世纪初建造完成的谷仓。这座谷物建筑的外墙由厚实的石灰石砖头建成，使用了金属地基板进行加固，地基板现在已经几乎丧失了原始功能，而是仅起到装饰的作用。该项目包括对建筑物的改造，将较矮的监控楼改建成饭店，将主楼改建成商业场所和舞蹈工作室。

整个项目完全是在原建筑的框架下组织建造的。一系列迹象表明，建筑的复建工程一点都不粗糙。例如，为了使新餐厅有足够大的空间，建筑师将屋顶抬高了一米，保留了屋顶原有的形状，并在屋顶抬高后露出的缝隙上安装了新的窗户。在这个建筑项目中，建筑师决定充分利用原有历史建筑的恢弘感，不开一扇窗户，只靠天窗来获取光线照明。建筑师经常使用视觉模拟方法，从外部看，屋顶几乎是唯一可见的改建部分，与之前提到的地基板紧密相连。

与之前提及的项目相似的是，建筑师总是利用一些小细节来调和原始建筑与修复建筑之间的冲突，例如，在建筑底层保留一排醒目的混凝土储料槽，以提醒人们勿忘建筑物的历史。改建后的罗特曼谷物升降机仍是一个工业仓库，虽然有了新的居住者，但他们在建筑内部的活动与先前并无矛盾之处，在新建筑中显得自然协调。

南特高等艺术学院由富兰克林·阿兹建筑设计公司于2017年设计而成，其原址为阿尔斯通仓库（66页）。该项目将原建筑整体拆毁，

space. The outer blocks are completely renewed with mid-tech grey curtain walls that claim the publicness of the building and wipe out its previous life with different facade geometries that emphasize different functions and a cantilevering glass block that faces the surrounding landscape.

For this reason, the Park shows two different aspects. On the street side, its glistening skin conveys an idea of bright, metallic and anodyne process of innovation. On the inside and in the courtyard, the building features a particular red brick old-fashioned industrial facade characterised by a red corten roof extension. Featured by sharp geometries, the roof tends to celebrate the idea of substitution, the old and heavy versus the new and light. Except from the chromatic and geometric peculiarity of the preserved parts, in all the building there seems to be no friction between the two phases. Every detail is carefully designed, whilst every element of neglect (e.g. rust, missing edges) is removed.

The Rotermann Grain Elevator (p.26) by KOKO Architects is located near the historic centre of Tallin, Estonia. The original building was completed at the beginning of the 20th century and served as a grain depot. The Rotermann building's walls are made of thick limestone bricks and reinforced with metal anchor plates, which have nearly ceased their function and now mainly serve as a decorative motif. The project consists of the adaptation of the building to house a restaurant in the short supervisory building and a series of business premises and dance studios in the main building.

The programme is organised entirely inside the old shell. A sequence of visual hints suggests it has been restored but never coarsely. To make space for the restaurant, for instance, the roof has been raised by about one meter, preserving the original shape and providing new windows in the gap created. In this project, the architects decided to exploit the monumentality of the historic building without opening any window and relying only on the skylights as a source of illumination. The visual approach used by the designers is often mimetic, the roofs being almost the only visible parts of the renovation works from the outside, in a close relationship with the aforementioned metal plates.

In a similar way, the conflicts between the old building and the restoration project are always mediated by an ap-

只保留并修复了其醒目的金属框架，用以支撑新建筑。从外部看，整个建筑就像一个巨大的锯齿状聚碳酸酯仓库，乍看起来不像是修复重建工程。在设计新学校的时候，保留下来的钢框架给了建筑师很大的自由发挥空间。有趣的是，聚碳酸酯挑篷的最前面的部分被空了出来，以便于人们进入建筑。不过从这点看来，建筑钢结构的工业外观也变得一目了然了。

然而，尽管在我们先前提到的项目中，新旧建筑之间有着清楚的界限，但在南特项目中，建筑师并没有颠覆建筑物原本的设计，而仅仅是赋予了其新的功能。学校内部的设计就体现了建筑师的这一设计策略。在学校建筑内部，很多地方都是光秃秃的，与工业环境很相像。学校建筑的内部空间与其外壳是分离的，因为整座建筑都围绕着金属框架而建，新旧建设之间并没有反差或冲突。在南特，高等艺术学院的工业遗迹虽然代表了一种乌托邦式的理想主义，但它也不失为对当代城市问题的一种解决方法，创造出了一种受保护的却又不失延续性的公共空间。

创新动力工厂（36页）是由Atelier Van Berlo建筑设计工作室、Eugelink建筑设计事务所和德贝弗建筑设计事务所联合设计的。创新动力工场的设计理念与北京文化创新工场相似，二者都认为畜牧业和重工业必须让位于非物质的重要新型工业。该项目原为飞利浦工业的发电厂，始建于20世纪初期后半段，是埃因霍温的标志性建筑。与罗特曼谷物升降机相似的是，该项目显然是想保留原有建筑的功能，对其外立面加以改造并增强建筑内部的可见性。该项目将废弃旧电厂改造成新的办公楼，在这个改造过程中，新旧之间的摩擦和分歧最终都被接受并彰显出来。旧建筑中的混凝土梁柱、储料槽和锈迹斑斑的钢架，与新建筑光亮的地板和闪光的玻璃墙面和谐共存。窗框仿照预制柱上混凝土梁枕的样式制成，柱子底部正常的磨损也展露出来，彰显着过往的荣光。在这项改建工程中，旧工业以及19世纪重工业概念被设计师用作装饰物，来装点迷人的新式工业建筑。

proach that tends to the minimum, leaving, for example, a sequence of impressive concrete hoppers at the ground floor as a reminder of the building's past. The Rotermann Grain Elevator remains an industrial depot, inhabited now by new life, whose activity does not conflict with the old form and naturally fills the space.

The higher school of fine arts of Nantes, formerly the "Alstom Warehouses" (p.66), was designed by Franklin Azzi Architecture in 2017. The project concerns the complete demolition of the old building, with the exception of the impressive metal frame that has been restored to support the new envelope. From the outside, the building appears as a huge polycarbonate saw-toothed warehouse, which is not immediately recognizable as a restoration work. The presence of the steel frame has allowed the architects a high degree of freedom in designing the new school. Interestingly, the first part of the polycarbonate canopy has been left clear to favour the entrance. From this point, though, the industrial appearance of the steel frame becomes evident.

However, whilst in the previous projects we noticed a clear-cut separation between the old and the new, in this case the building simply continues its life with a renewed function. This choice is displayed by the architect in the visual appearance of the interiors of the school, which are often left bare and resemble an industrial environment. The inner volumes are left separate from the shell, and because the whole compound is built around the metal frame, there is no contrast or friction between the two phases. In Nantes, the industrial remains of the higher school of fine arts, by virtue of their scale, represent a utopian but nonetheless possible solution of contemporary urban problems, creating a protected but still continuous public space.

Innovation Powerhouse (p.36) was designed by Atelier van Berlo, Eugelink Architectuur and De Bever Architecten. In a similar fashion to the Beijing Cultural Innovation Park, the Innovation Powerhouse serves the idea that the brute and heavy industries of the past must give way to a new model of industry, related to the immaterial and the relational. Built in the second half of the 1900s, the original building was an iconic landmark of Eindhoven and served as a power plant for the Philips industries. Similarly to the Rotermann Grain Elevator, this project apparently aimed at inhabiting the old power plant, yet worked on its public facade and sought to enhance its visibility. Every friction in

1. Debord, G. The society of the spectacle, 1994, Zone Books, NY, p.10

本文介绍的最后一个项目是EFFEKT建筑设计事务所设计的维堡街头圣地（82页）。该项目将工业仓库改建成了集滑冰、跑酷等街头运动为一体的室内活动场所。在该建筑案例中，由预制板混凝土建成的旧仓库被套上了由细钢柱支撑玻璃板镶嵌的新外壳，原仓库大厅双排通道两侧的密闭空间也被扩大了——其中一条通道两侧的房间用作了办公室和服务台，另一条通道一侧则建造了滑冰场，与室外的滑冰区相连。该建筑设计颠覆了富兰克林·阿兹建筑设计公司对南特高等艺术学院项目的设计：建筑内部保留了光秃秃的旧结构，而带有岁月痕迹的旧外壳也被清晰地展示了出来，但是因为旧建筑大厅设计简单，EFFEKT因此比Van Berlo等其他设计团队更容易化解新旧建筑之间的矛盾。所有的矛盾都被化解了，旧建筑的衰败也在钢筋和胶合板新架构中表露无疑，但是即便如此，建筑中每个被废弃的部分都被展现出来并中和掉了，这使得旧工业仓库看起来像是郊区环境的投影。

这些建筑项目阐述了传统生产线那严谨的空间结构是如何被打破、暴露并展示出来的，并被更具流动性、更灵活、视觉传达更为优异的空间所取代，这也反映出了当今经济朝着自动化和非物质化转型的趋势。以南特高等艺术学院和维堡街头圣地为例，前者保留了原建筑的金属框架，而后者则采用新建筑将旧建筑包裹其中。这些案例使我们了解到，19世纪现代社会的特点是如何适用在当今社会之中的。从某种角度来看，旧工业建筑是动乱却又辉煌的旧时代的标志，对其中的某些部分无需进行大幅度的改动就可以很容易满足现代建筑的需求。

the transition from the old neglected power plant to the new office building is here accepted and emphasized. The old concrete beams and columns, the hoppers and the rusty steel frames are juxtaposed with the polished floors and the shiny glass facades. The window frames follow the concrete corbels on the prefab columns, and the normal wear and tear signs at the bases of the columns are exposed with a sense of pride. The old industry, and the idea of the 19th century heavy industry, is displayed as an ornament for the new, glamorous forms of production.

The last project of this section is Streetmekka Viborg (p.82), from EFFEKT, which consists of the adaptation of an industrial warehouse into an indoor arena for street sports such as skating or parkour. In this case, the old prefabricated concrete warehouse, which consists of a two-aisle hall, has been enclosed into a new shell made of slender steel columns and glass panels, which expands the enclosed space on two new aisles – one accommodating offices and services, the other one hosting a skate pool – near to the skate area on the outside. This approach overturns the one adopted by Franklin Azzi Architecture in Nantes: the bare old structure is left inside and the old shell, the sign of its age, is clearly displayed, but the simplicity of the pre-existing hall allows for a subtler approach in the resolution of the conflicts between the old and the new than the one shown by Atelier Van Berlo et al. at Innovation Powerhouse. All conflicts are resolved and the decay of the old is displayed in the frame of the new steel and plywood additions, but nevertheless, every bit of abandon is framed and neutralised, making the old warehouse look like a mere scenography of a suburban environment.

These projects show how rigid spatial configurations of traditional assembly line are disrupted, exposed and celebrated, replaced with other more fluid, flexible and visually communicating spaces, reflecting today's economic shift toward automation and dematerialization. However, through few cases such as the higher school of fine arts of Nantes, whereby only the metal frame is preserved, or the Streetmekka, where the new encases the old, we can appreciate how some of the features of 19th century modern societies are still appropriate in today's world. Somehow, industrial archaeology shows the signs of a troubled and glorious past, of which a part can easily accommodate today's needs without radically modifying its original role.

罗特曼谷物升降机
Rotermann Grain Elevator

KOKO Architects

罗特曼街区位于爱沙尼亚塔林老城区的历史性重要位置，处于旧城区、港口和维鲁广场之间。自19世纪开始，维鲁广场便已经成为通向塔尔图、纳瓦和帕尔努的路线交叉口，成为塔林市的中心要塞。罗特曼街区历史建筑林立，其密度几乎可以和旧城区相媲美。罗特曼工厂成立于1829年，其创始人克里斯蒂安·亚伯拉罕·罗特曼引领了该密集工业区的发展。这些年来，坐落其中的工厂和贸易公司历经兴衰。这里的建筑在苏联时期曾遭到重创，又在其后的时间里逐渐衰败乃至荒废。1979年，这个破败的街区成为安德烈·塔尔科夫斯基的电影《追踪者》的取景地。2001年，爱沙尼亚国家遗产保护局将此地列为历史遗产保护地。至此，这些被废弃的工业建筑终于在群楼环绕的当代城市里找到了自己的位置，与当代建筑和平共处。

该街区最壮观的建筑是一栋1904年建成、位于Hobujaama街的谷物升降机，其前方建筑被规划为一座饭店。这座狭窄的建筑长一百多米，其长立面不设开窗，而用于加固建筑的铁箍穿透了石灰岩外立面。当谷仓中的谷物膨胀时，这些铁箍也有助于保持建筑外墙的完整性。这些从内伸出来的铁箍，令该建筑外立面一眼望去像是一件布满了纽扣的旧外套。该建筑屋顶比原先的高度抬起一米，如同悬浮着，其目的是将充足的自然光引入室内，并创造出可以使用的二层空间。

临街的墙体上有一些洞口，这些洞口曾在不同历史时期被封闭，建筑底层因此被用作商务办公场所。这些房间中留有老旧的谷物储料器，悬于天花板之上。建筑底部中间有一条拱廊，将空间一分为二，形成了进入内大街的入口，通往街区中心。位于建筑内部的舞蹈工作室不设开窗，而位于屋顶阁楼的办公室不仅光线充足，还可以从这里欣赏到老城区的迷人景致。

AED1904
RNV1930

南立面 south elevation

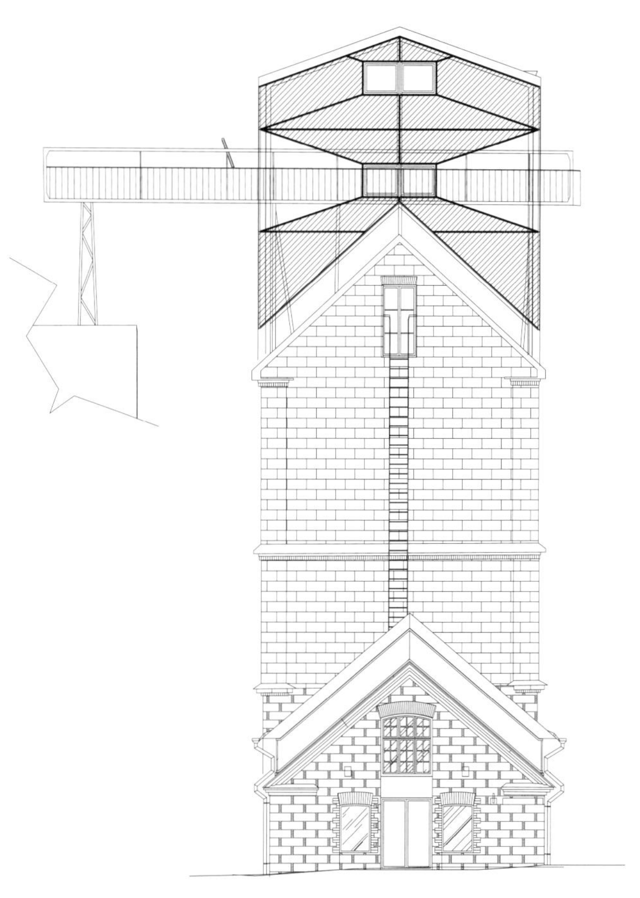

北立面 north elevation

The Rotermann quarter is in a historically important location in the heart of Tallinn – between the Old Town, the harbour and Viru Square. The roads to Tartu, Narva and Pärnu have already been intersecting on Viru Square in the 19th century, making it Tallinn's official central point. The Rotermann quarter is packed with historical buildings almost as densely as the Old Town. Christian Abraham Rotermann, the owner of the enterprise Rotermann Factories, established in 1829, initiated the development of the compact industrial district. Until today, industry and trade in the quarter have undergone ups and downs. The Soviet years wrecked the buildings and during the uncertain years that followed the buildings became dilapidated so that repairs seemed impossible. In 1979, the decaying district became the set for Andrei Tarkovsky's world famous movie "Stalker". The National Heritage Board designated the Rotermann quarter historically valuable in 2001, and so the old industrial buildings should be given a new function to coexist peacefully with high-quality contemporary architecture.

One of the most spectacular buildings in the Rotermann

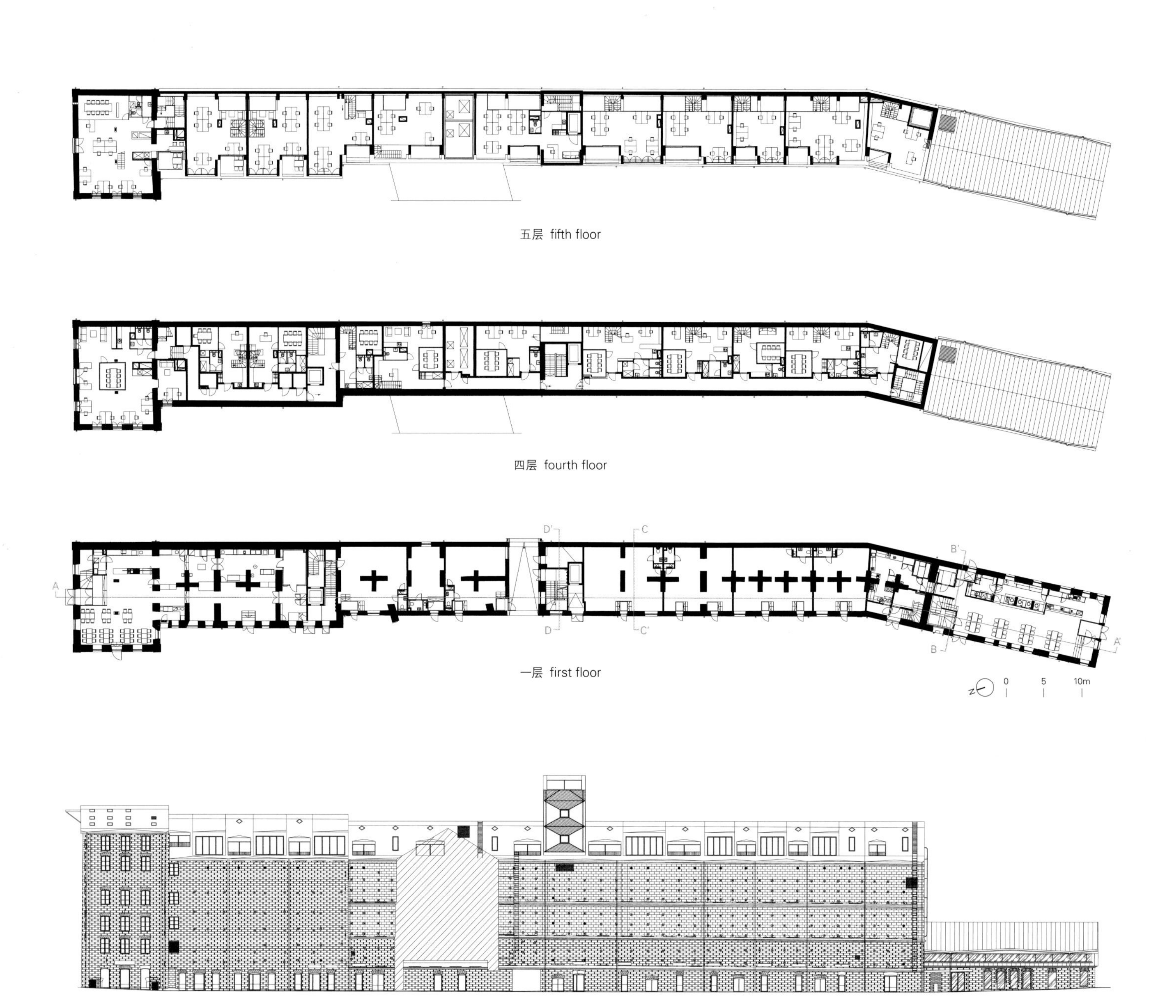

五层 fifth floor

四层 fourth floor

一层 first floor

西立面 west elevation

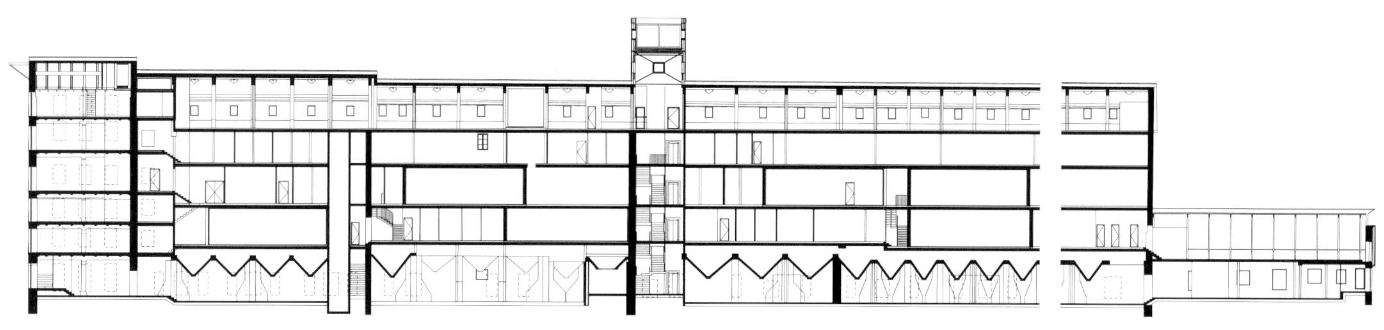

A–A' 剖面图 section A-A'

AED1904
RNV193

quarter, the grain elevator was completed in 1904 on Hobujaama Street, behind the historic supervisory building for a restaurant. The narrow building is over 100 meters long. The longer sides have no windows, but instead, the limestone facade of the building is accentuated by metal straps that reinforce the wall. The straps had the purpose of keeping the grain elevator walls intact even when the grain expanded. The wall is packed with metal details, like a useful old coat covered in buttons. The roof of the building has been raised by one meter, thus appearing to hover. The aim was to let natural light enter and make it possible to utilize the second floor.

Since the inner street side has openings that have been walled shut at various periods, the ground floor of the building houses business premises. The interiors of these rooms have preserved the old grain hoppers hanging from the ceilings. An arcade that crosses the middle part of the building on the ground floor divides the space and creates an entrance to the inner street leading towards the center of the district. Dance studios are accommodated on the floor without windows and the attic provides offices with skylights that look out across the district and the Old Town.

项目名称：Rotermann Grain Elevator
地点：Tallinn, Estonia
建筑师：KOKO architects - Raivo Kotov, Olga Batuhtina, Andrus Kõresaar
室内设计师：Raili Paling, Kadri Kaldam
客户：Rotermann City OÜ
面积：5,600m²
委托时间：2007
竣工时间：2016
摄影师：©Tonu Tunnel (courtesy of the architect)

B-B' 剖面图
section B-B'

C-C' 剖面图
section C-C'

D-D' 剖面图
section D-D'

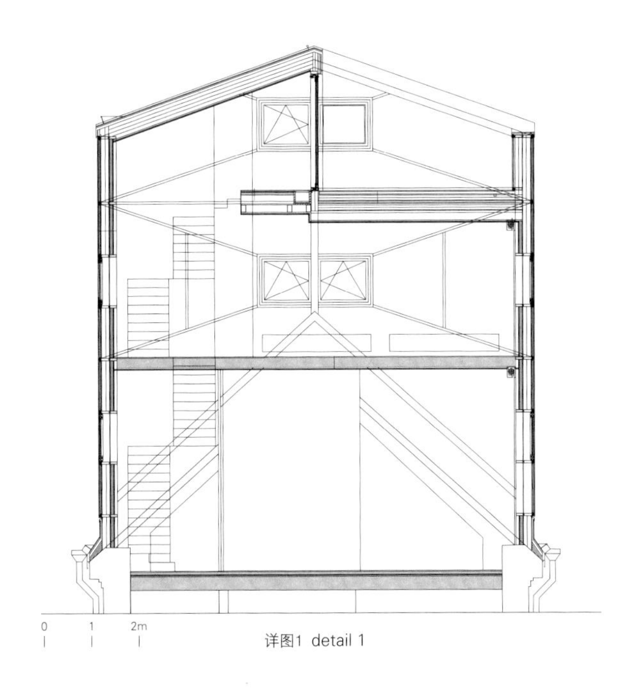

详图1 detail 1

创新动力工场

Innovation Powerhouse

Atelier van Berlo + Eugelink Architectuur + De Bever Architecten

Atelier Van Berlo建筑事务所、Eugelink建筑设计事务所和De Bever建筑设计事务所将一座旧发电厂改造成了一个容纳多间公司的创新动力工场。原建筑是一座原始的重工业综合体，用途单一且建造粗糙，其大小完全不适合用作办公场所。而建筑师将其改造成了一个舒适、通透、宽敞、新颖而鲜亮并且充满年轻活力的工作场所。

创新动力工场是创新产业发展的一个平台，是一座汇集创造力的超现代主义建筑。各类公司在此聚集，思维的火花在此碰撞，激发出创新和灵感。该废弃的发电厂坐落于Strijp-T区，是埃因霍温的标志性建筑。该电厂共分四个阶段建成，建设工程从1953年持续到1972年，曾使用煤、天然气以及后来的石油等原料为飞利浦工厂发电。

该改造项目最早由Atelier van Berlo建筑事务所发起。作为一家以设计为导向的建筑事务所，他们在为自己的公司寻找新总部时，注意到了这座建筑。在这座建筑的改造设计中，他们不仅将这座建筑当作其新总部的所在地，还将它设想为一个独特的工作空间、一个专注于合作的开放生态系统。最终，它将成为一个生态平台，供各类创新公司、大学院校、合作伙伴、客户和访客携手努力，致力创新，共塑未来。

该设计的两个主要出发点是保持建筑原有的品质，并融入开放创新的理念。因此，设计方案必须将这两种想法融合。建筑师希望在创新动力工场落户的公司们能够时常碰面，相互汲取灵感，将这座废弃的建筑打造成一个充满活力的中心。

为了实现这个目标，建筑师在建筑内部创造了一个清晰的轮廓，一条直线从原始建筑的中心穿过，将其一分为二。5m宽的原始中心支柱仍然矗立其上，在支柱28m高的地方，种植斜槽仍然悬挂其上。中央支柱的旁边是横跨整个屋顶的天窗，天窗引入的光线照亮了原本黑暗的中心地带，同时也展示出了古旧的重型混凝土结构的庞大规模。此外，因为天窗的存在，用户在外立面就可以一睹建筑中心支柱的高大。

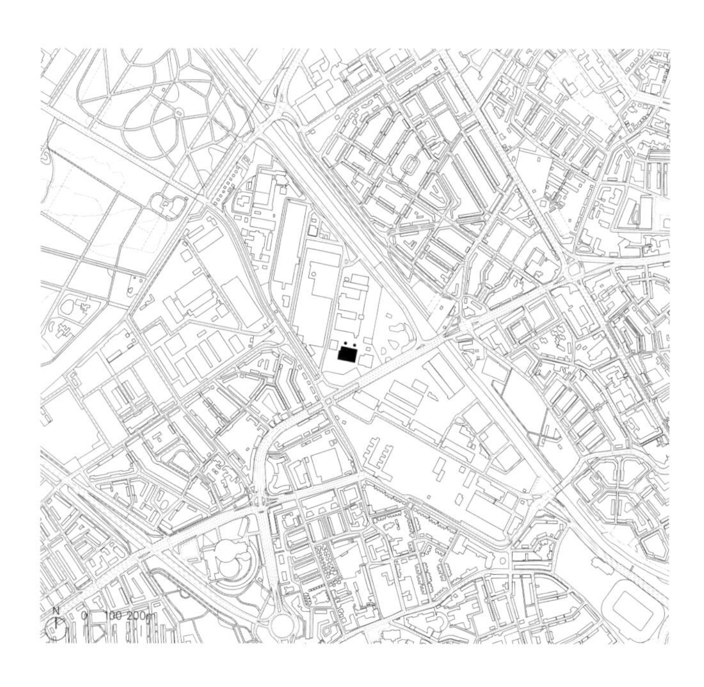

INNOVATION POWERHOUSE

由于发电模式不断变化，原建设一直没能彻底完工。例如，种植斜槽已经不再需要了，5m宽的高层结构仍未完工。建筑师从原始图纸上汲取灵感，在中央高柱上增建了一座垂直钢铁花园，使建筑恢复了设计之初的对称美。这个垂直花园中设有会议室、玻璃电梯和紧急出口，遵循了原有建筑的线条和节奏。开放、透明的绿色外观暗示着可持续的绿色能源生产。建筑师保留了原有建筑的特色，引入了简约的设计细节和现代材料，使这座壮观的建筑重获新生。

还有一些其他的改动，例如，在办公室和主入口开设小窗户采光，这些微小的改动保持了这座宏伟建筑的工业感。在改造过程中，建筑师也非常注重可持续性。通过使用HR+玻璃、太阳能电池板、新隔离墙和屋顶，这座建筑如今获得了A+能源标签。

Atelier van Berlo, Eugelink Architectuur, and De Bever Architecten transformed a former power plant into the multi-tenant Innovation Powerhouse. A transformation from a pure, brute, raw and heavy industrial complex - in size and volume completely unsuitable for office use - into a comfortable, breathing, spacious, fresh and young work environment.

The Innovation Powerhouse is a stage for the innovative industries – a center of creativity, a hyper-modern building where different companies meet, inspire and stimulate innovation. Situated on Strijp-T, the out of use power plant was an iconic building, a landmark of Eindhoven. Built in four stages, from 1953 till 1972, the power plant used to provide power for Philips factories through coal, gas, and later oil.

The design-driven innovation agency, Atelier van Berlo initiated the transformation, while in search for their own new headquarters. They envisioned a unique working space, an open ecosystem focused on collaboration, something more than just their new headquarters. In the end, it would be an ecosystem for different innovative companies, universities, partners, clients, and visitors to work together and create the innovations that shape the future.

The two main points of departure for the design were to maintain the building's innate architectural qualities and to incorporate the vision of open innovation. Thus, the design

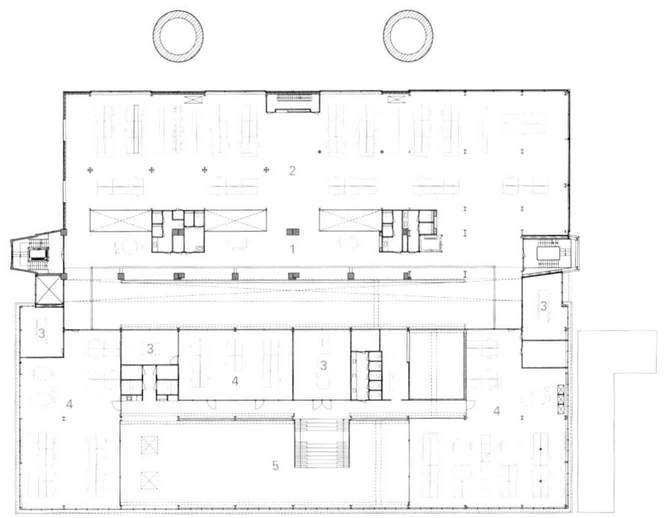

1. 中央通道 2. F出租户型 3. E出租户型——会议室 4. E出租户型——工作区 5. E出租户型——中庭
1. central hallway 2. tenant F 3. tenant E-meeting room
4. tenant E-workplaces 5. tenant E-atrium
三层 third floor

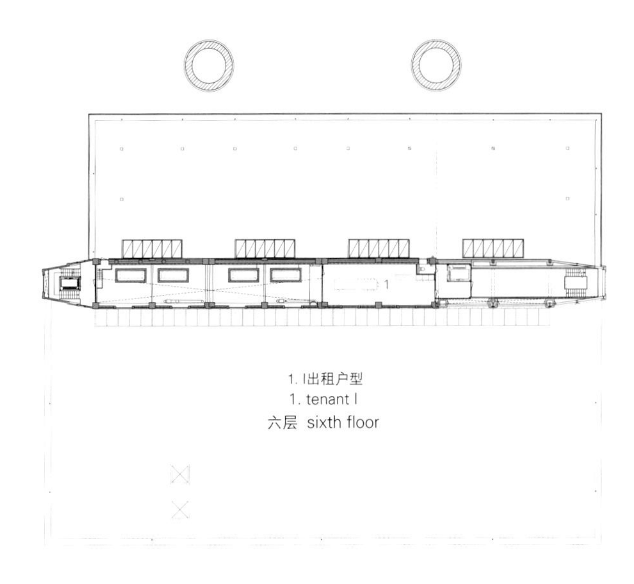

1. I出租户型
1. tenant I
六层 sixth floor

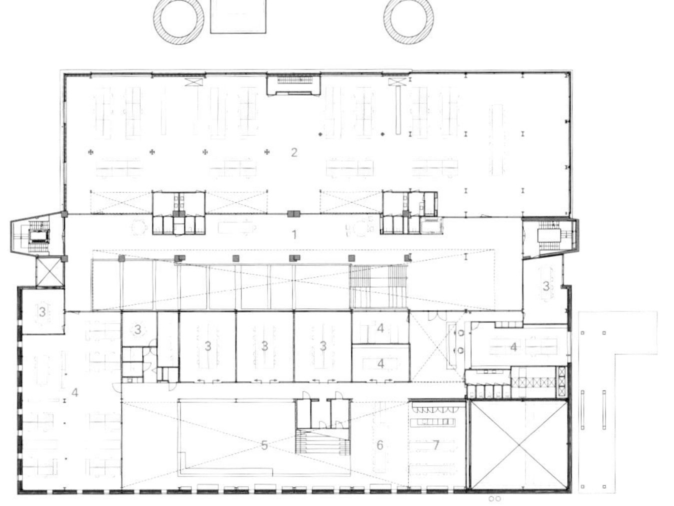

1. 中央通道 2. D出租户型 3. E出租户型——会议室 4. E出租户型——工作区
5. E出租户型——中庭 6. E出租户型——咖啡吧 7. E出租户型——厨房
1. central hallway 2. tenant D 3. tenant E-meeting room 4. tenant E-workplaces
5. tenant E-atrium 6. tenant E-coffee bar 7. tenant E-kitchen
二层 second floor

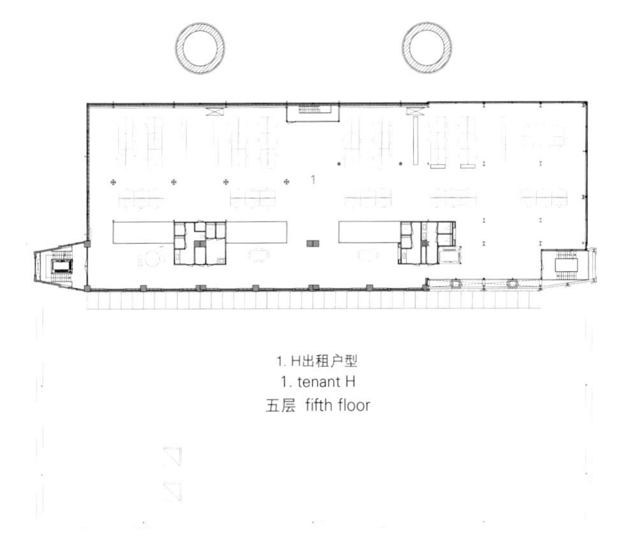

1. H出租户型
1. tenant H
五层 fifth floor

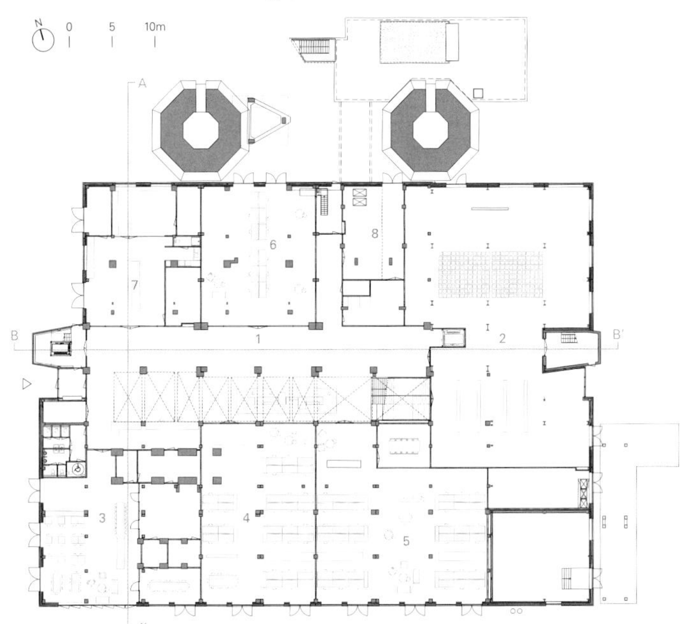

1. 中央通道 2. 会议厅 3. 餐厅/酒吧 4. A出租户型
5. B出租户型 6. C出租户型 7. 工作坊 8. 系统&设备间
1. central hallway 2. conference hall 3. restaurant/bar 4. tenant A
5. tenant B 6. tenant C 7. workshop 8. systems&services
一层 first floor

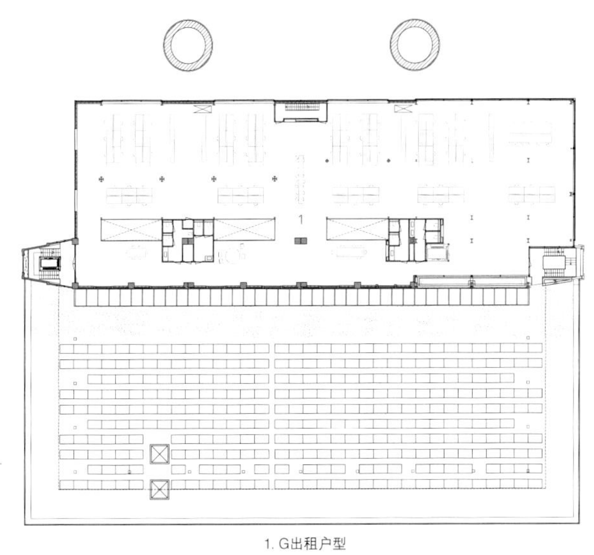

1. G出租户型
1. tenant G
四层 fourth floor

ABUS

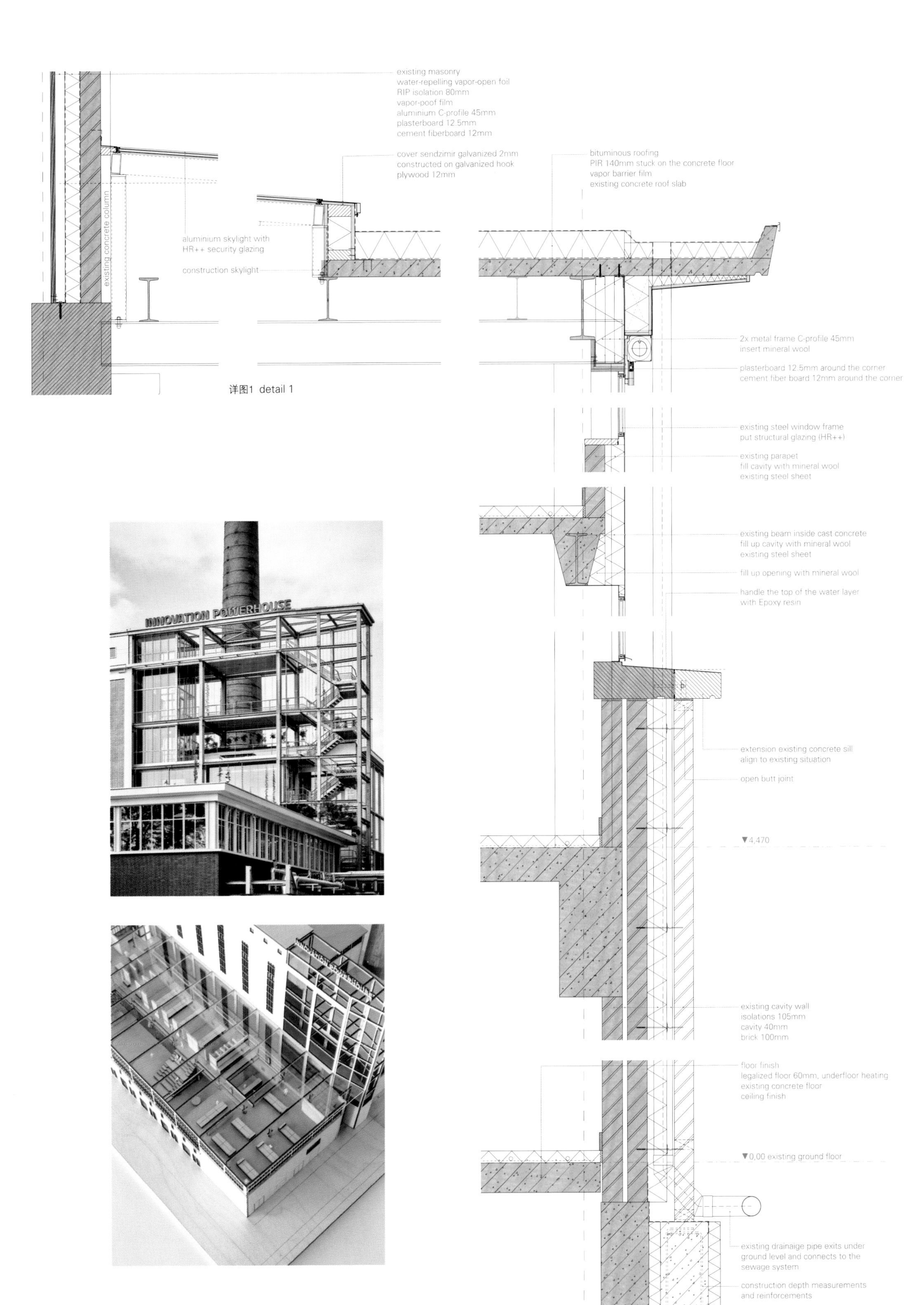

详图1 detail 1

详图2 detail 2

blending these two ideas was essential. The architects wanted the companies to constantly meet, to effortlessly inspire one another, which will translate the abandoned structure into a vibrant center.
To achieve this, the architect created a clear-cut in the building, a straight line through the original heart spanning among others the original central backbone, a 5-meter wide high-rise where the cole chutes still hang at 28-meter height. Next to this central backbone, the roof is opened up by a skylight spanning the full length, bringing light into the otherwise dark center of the building and revealing the magnitude of the old heavy concrete structure. In addition, this skylight gives a peek at the facade, allowing users to experience the sheer height of the midpoint.
Due to a change of trends in power generation, the original design was never finished. For example, the cole chutes were no longer in need, and the 5-meter wide high-rise remained unfinished. Inspired by the original drawings, the architects added a steel vertical garden to the central high-rise, giving the building its originally intended symmetrical look. This vertical garden with meeting rooms, a glass elevator, and emergency exit follows the lines and rhythm of the original architecture. The open, transparent, and green appearance hints at a sustainable green energy production. Preserving the characteristic features of the original building and introducing the minimalistic detailing and modern materials, the architects clearly gave a new life to the impressive building.
The other interventions, such as the small windows for light in the offices and the main entrance, are kept small, to ensure that the monumental industrial character of the building stays visible. Within the renovation, there was also a great attention for sustainability. With HR+ glass, solar panels, newly isolated walls and roofs, the building now obtained an A+ energy label.

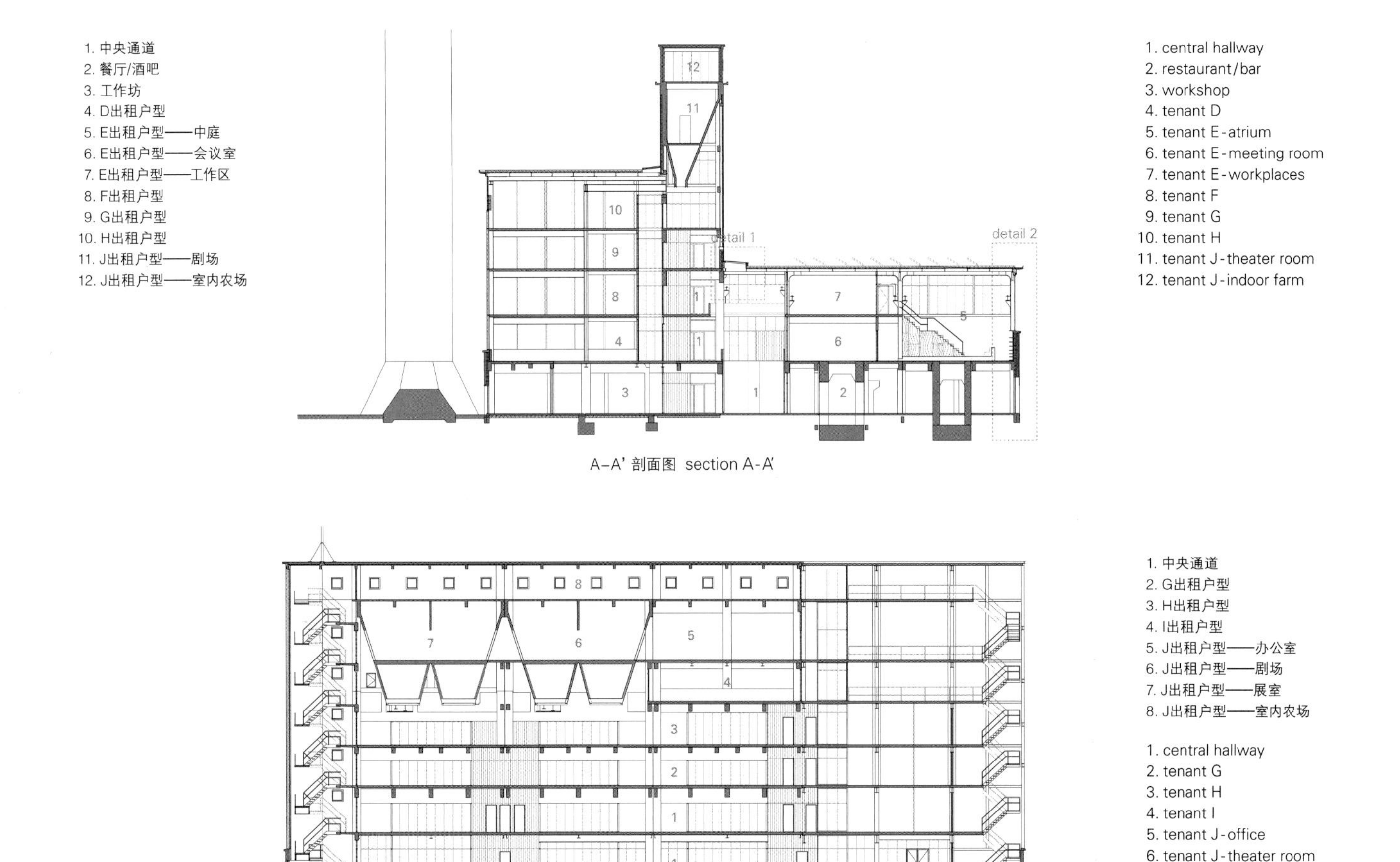

A–A' 剖面图 section A-A'

B–B' 剖面图 section B-B'

北京文化创新工场

Beijing Cultural Innovation Park

Cobblestone

该项目位于北京中轴线南端的国家新媒体产业园区内，距天安门约16km。该园区最早是建立于上世纪90年代的传统工业园区，2005年成了以新媒体产业为主导的专业产业集聚区。依据2013年材料控制性规划，园区内既有的80多家工业企业将转型为以文创、服务、信息、数字产业、商务为主要内容的高科技产业。而该项目"创新工场"的定位符合控制性规划和未来的开发需求。

所有产业地块的更新都伴随着产业和空间的双重更新。然而以产业为导向的内容组织往往是动态的和不固定的，因而建筑的空间形态需要持续反映潜在内容和未来投资的价值。在这一主旨下，"北京首兴创新工场"的改造设计目标是通过重组场地与建筑的内容、界面和肌理，建造出适用于潜在的文创领域功能的交流空间和场所。该建筑项目为科技和文创企业提供展示、交流、投资、金融服务、办公与休闲活动的支持平台与场所。这里的关键词是交流，"交流"即交叉和互动，这一理念可以渗透在空间界面、立面肌质、产业之间、信息交换、人际行为和情感之中，具有内容和形式的双重意义，与创新相契合。

场地组织

原先的工厂是采取的是封闭的格局，其流线由蓄煤、燃煤、传热过程发展而来。该项目根据未来的功能将场地进行切割的，沿街道外围设置设计了交易和交流性强的流线。同时，沿内院设置了文化活动和景观空间、配套设施等相对静态的流线。界面和流线的重塑和整合激发了场地的活力，实现了从个体到场所的转变。

建筑组织

为了实现这一转变，建筑师采用了两种方法重新定义空间与交互。其一是设计了叠加在西1号商品展示建筑顶部的厚实楼层（加层扩建）。中间以及水平方向的核心接触点是一个扩大的大厅和中庭；水平方向的结构在不同的高度分别为2号展厅、3号办公楼、5号步行通道（原煤炭运输机）和悬挑建筑的西侧。其二是内院周边如拼贴画一般分散布置的小体量单体建筑。从上方俯瞰，这些小建筑看起来就像一个整体，但只有从地面看，才能凸显这些小建筑之间的交互空间。两种设计方法使空间及场地之间产生了诸多交叉，在许多层面都体现出了交流的特点，完成了个体向场所的另一个转变。

美学策略

整个立面的设计语言都受到了技术表现主义和意大利理性主义的影响。在西侧，1号楼的入口大厅为钢结构，在三层设计了一个7m的外悬挑结构，用于展示和观景。西侧立面的钢悬臂与铝型材的波浪纹理在道路的拐角处交叉，形成了强烈的视觉冲突，增强了建筑形象的张力。在南侧，沿街建筑立面由起伏的钢铝玻璃和其他构件组成，体现出一种强大而又充满活力的材料特质。封闭场地的建筑室内以价格低廉的红砖贴面，与金属结合在一起既形成反差，又展现了协调之感，体现出温暖质朴的人文特征。2号楼原引风机房的烟囱被保留了下来，通过连续重复的山墙设计，形成具有鲜明理性主义特色的立面。下沉的美食广场顶部覆盖着一个连续的钢结构，不规则的圆锥形屋顶覆盖了平坦的耐候钢板表面。这一设计延续了现代与简约相结合的美学思想，充分展现了充满跳跃的内庭院的空间活力与丰富性。

这种当代与古典精神的融合，表现在改造现有的建筑时，创新与记忆、人文之间有趣的关系。这些基于内部和外部空间的特点以及空间形式语言和材料特点的多重反差和协调，表明在当前社会环境下，特别是在新媒体工业园区，沟通和互动作为一种行为和空间媒介，已然成为"事件"的催化剂。

1. 珠宝艺术中心 2. 艺术展览画廊 3. 商业及附属建筑 4. 自动扶梯走廊
5. 便利店 6. 啤酒吧 7. 带保温金属屋顶的下沉美食广场 8. 茶室 9. 咖啡吧

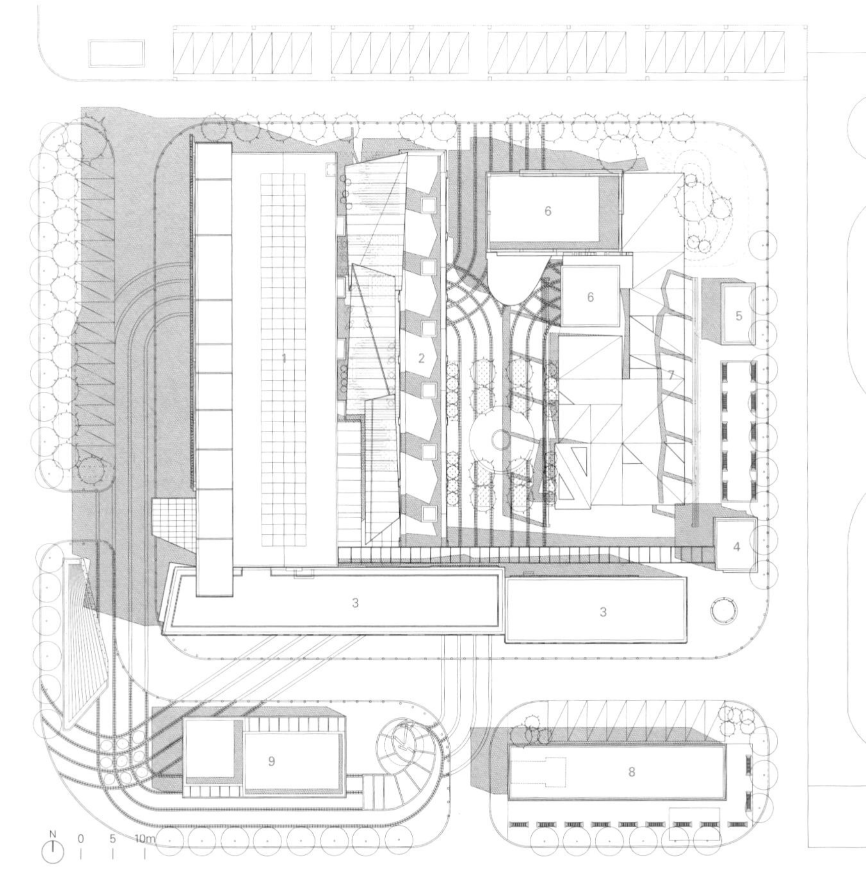

1. jewel artistic center 2. art exhibition gallery 3. commercial and auxiliary building 4. escalator corridor
5. convenience store 6. beer bar 7. sunken food court with insulated metal roof 8. tearoom 9. coffee bar

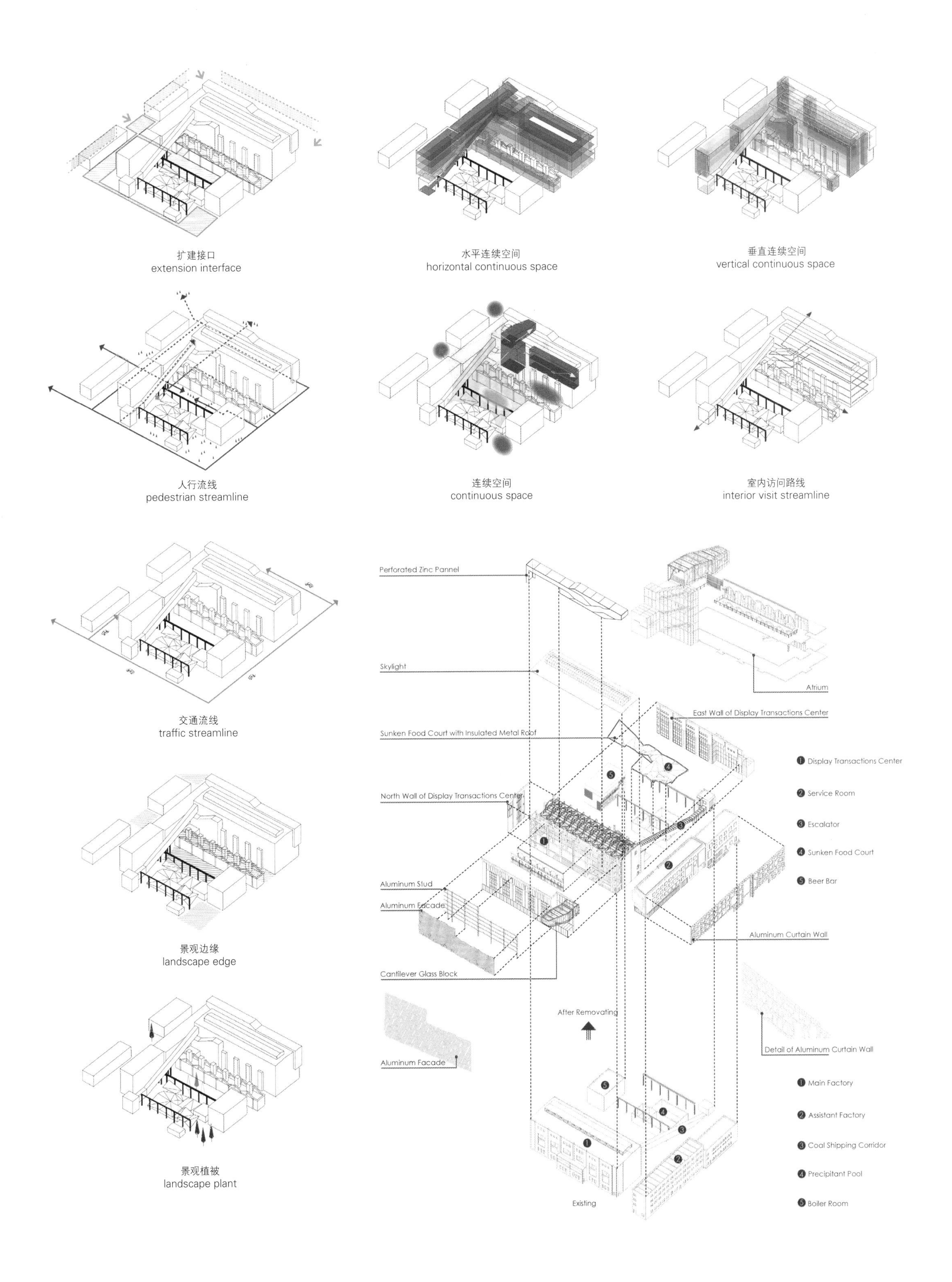

扩建接口
extension interface
水平连续空间
horizontal continuous space
垂直连续空间
vertical continuous space
人行流线
pedestrian streamline
连续空间
continuous space
室内访问路线
interior visit streamline
交通流线
traffic streamline
景观边缘
landscape edge
景观植被
landscape plant
Perforated Zinc Pannel
Skylight
Sunken Food Court with Insulated Metal Roof
North Wall of Display Transactions Center
Aluminum Stud
Aluminum Facade
Cantilever Glass Block
Aluminum Facade
Atrium
East Wall of Display Transactions Center
Aluminum Curtain Wall
Detail of Aluminum Curtain Wall
After Removating
Existing
1 Display Transactions Center
2 Service Room
3 Escalator
4 Sunken Food Court
5 Beer Bar
1 Main Factory
2 Assistant Factory
3 Coal Shipping Corridor
4 Precipitant Pool
5 Boiler Room

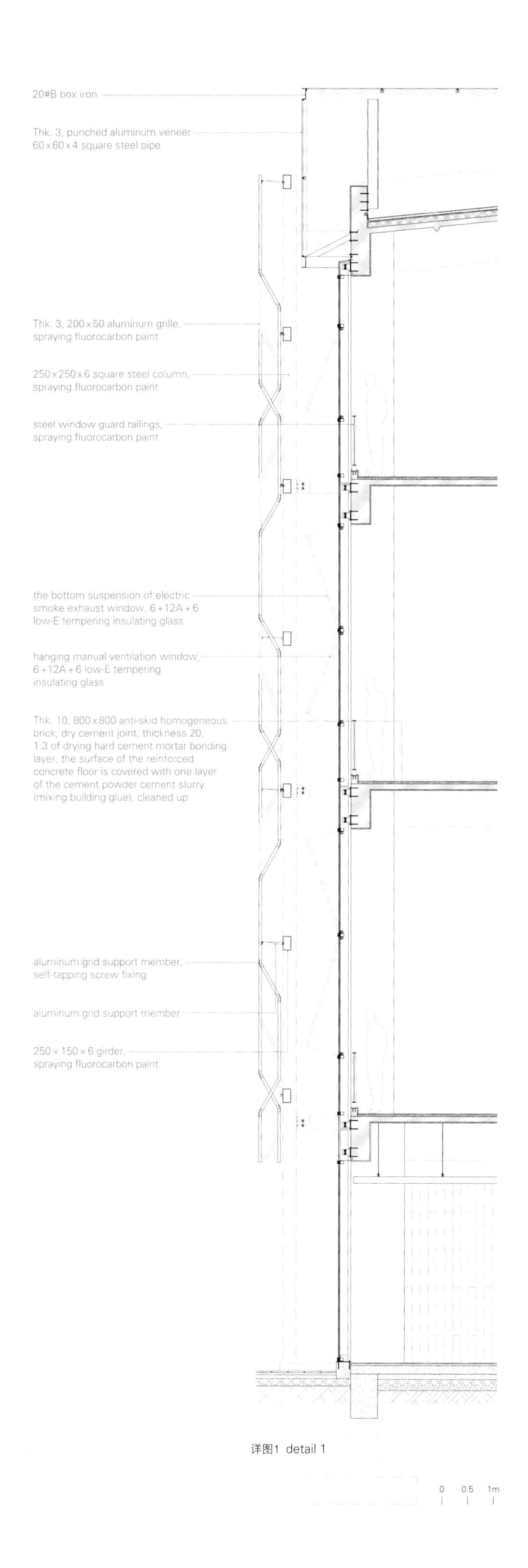

详图1 detail 1

The project is located in the China New Media Industrial Park at the southern end of the central axis of Beijing, about 16 km from Tian An Men. Originally built in the early 1990s as a traditional industrial park, it became a professional industrial agglomeration area dominated by new media industry in 2005. According to the material control plan 2013, more than 80 industrial enterprises in the park will be converted to cultural creative, services, information, digital industry, commerce as the main content of the high-tech. The project positioning "innovation works" conforms to the control regulations and future development needs.

The renewal of all industrial plots is accompanied by the dual renewal of industry content and space. However, the content organization under the industrial orientation is often dynamic and unfixed, so the spatial form needs to continuously reflect the value of the potential content and future investment. The goal of renovation and design of "Shouxing Innovation Workshop" was to build the communication spaces and venues which are adapted to potential domain functions by reorganizing the site and content, interface, and texture of building. It provides display, exchange, investment and financing services, office and leisure supporting platforms and venues for technological and cultural innovation enterprises. The keyword, "communication" – intersection and interaction, which can penetrate into the spatial interface, facade muscle, industries, information exchange, interpersonal behavior and emotion, had the dual meaning of content and form, and innovation fits.

Site Organization

The original factory had a closed pattern, with streamlines developed by coal storage, coal combustion, and heat transfer process. Cutting the site according to its future functions, highly transactional and communicating streamlines is set along the open periphery. Relatively static streamlines of cultural activities, landscape space, catering facilities, etc., are set along the garth. Interfaces and streamlines are rebuilt and integrated, which inspire the vitality of the site and achieve the transformation from the individual to the place.

Architectural Organization

There were two ways to redefine space of the interface for such transformation. One is a bulky layer stacked on the top of the West 1 trade show building. The core point of contact in the middle and horizontal direction is an enlarged hall

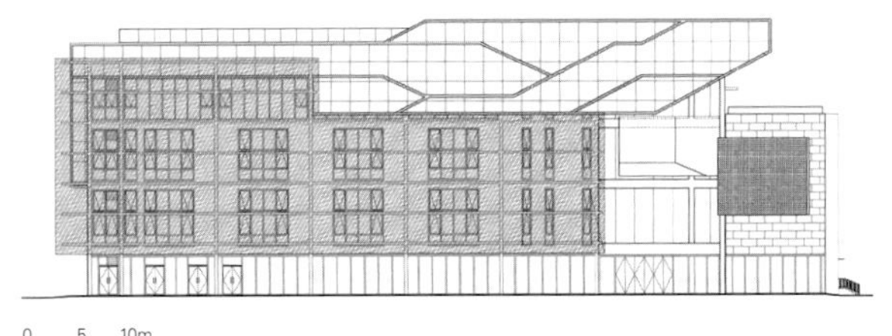

西立面 west elevation

南立面 south elevation

东立面 east elevation

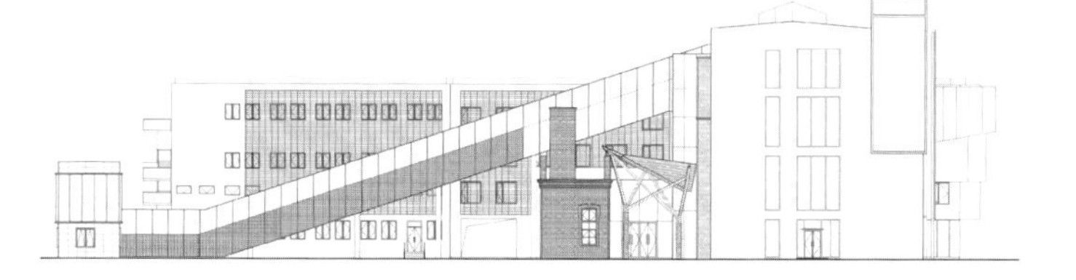

北立面 north elevation

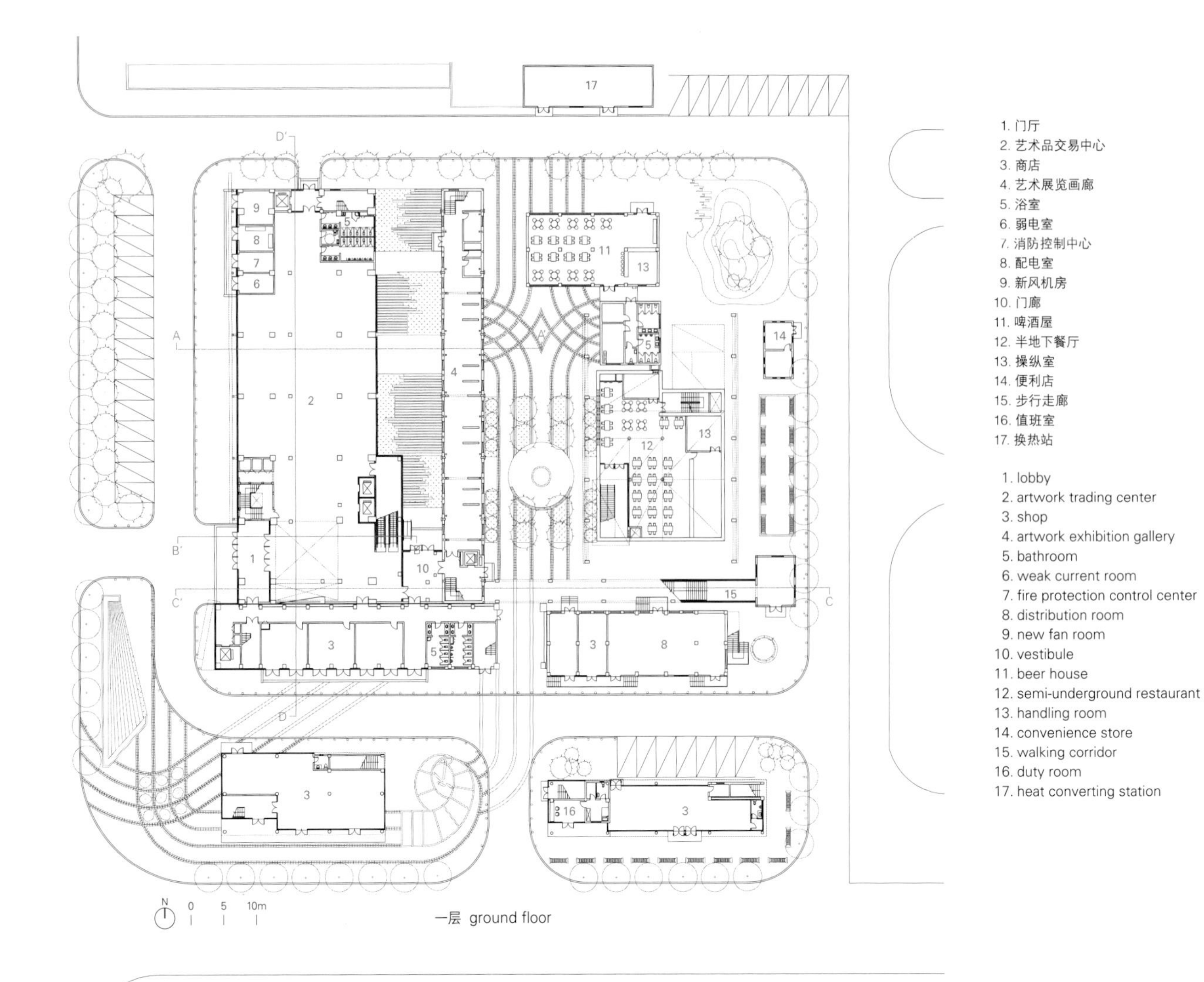

一层 ground floor

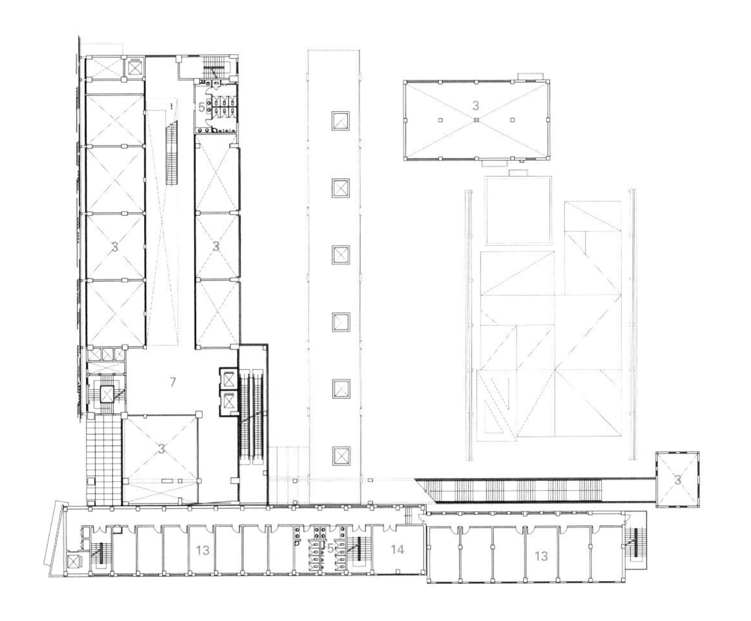

夹层 mezzanine

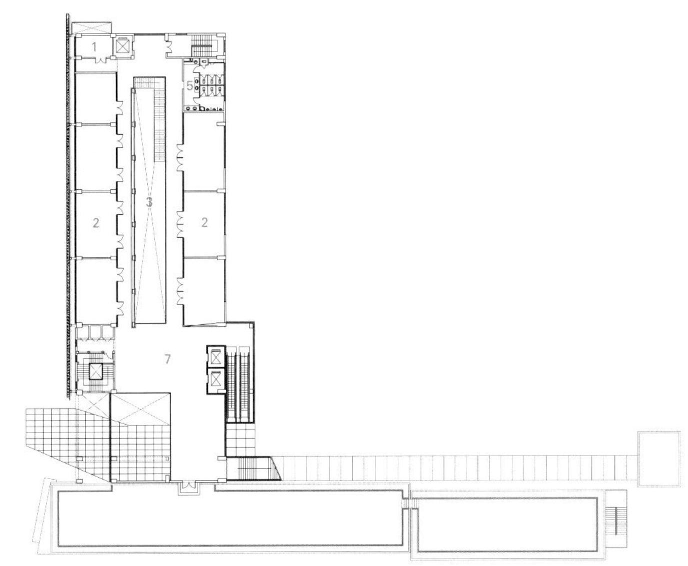

三层 third floor

1. 新风机房	6. 弱电室	11. 茶室
2. 艺术品交易中心	7. 走廊	12. 操纵室
3. 高架	8. 门廊	13. 办公管理室
4. 艺术展览画廊	9. 机房	14. 会议室
5. 浴室	10. 咖啡吧	15. 观景平台

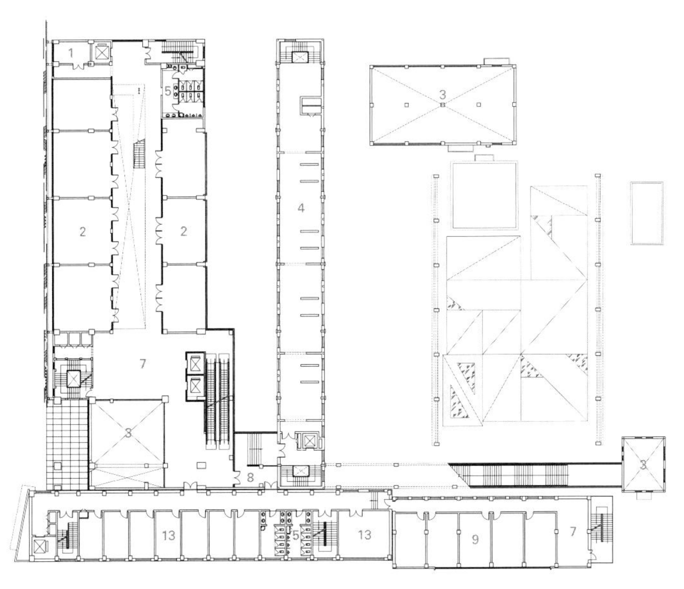

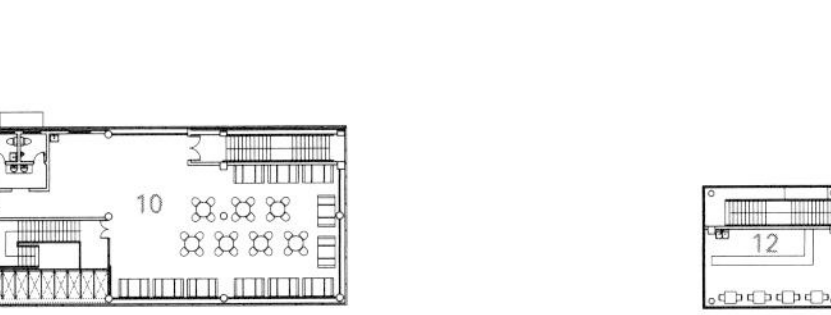

一层 first floor

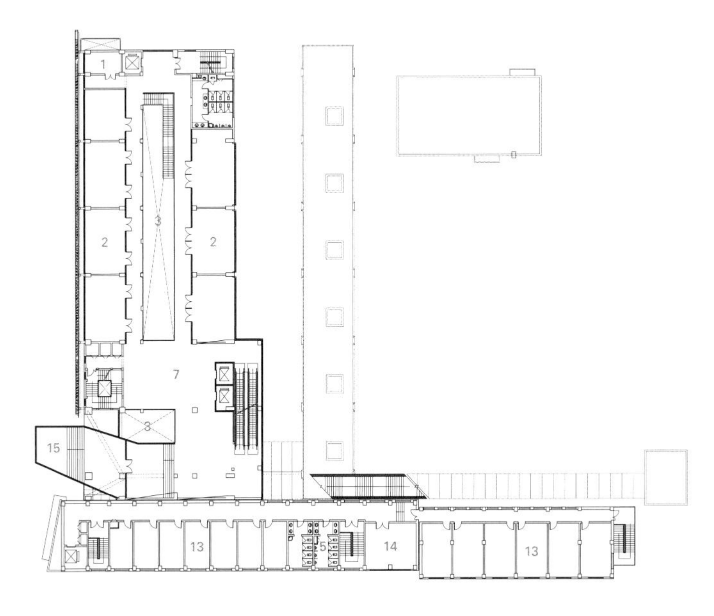

二层 second floor

1. new fan room	6. weak current room	11. tearoom
2. artwork trading center	7. hallway	12. handling room
3. overhead	8. vestibule	13. office management room
4. artwork exhibition gallery	9. equipment room	14. meeting room
5. bathroom	10. cafe bar	15. viewing platfrom

and atrium; the horizontal direction is at different elevations with exhibition hall No. 2, office building No. 3, pedestrian corridor No. 5(original coal conveyer) and the west side of the cantilevered architectural link. The other one is the collage of figure ground shaped from decentralized smaller blocks around the courtyard. Together they seem like one building from above, but the interactive space formed by these blocks is highlighted only from the street level. These two paths producing a lot of intersections between spaces and the site present the characteristics of communication at many points and surfaces, thus completing another transformation from isolated individuals to places.

Asthetic Strategy

The design language of the whole facade is influenced by technical expressionism and Italian rationalism. On the west side, the entrance hall of building No.1 is designed as a steel structure with an outward cantilever of 7m on the third floor for display and view. The steel cantilever and the wavy texture of the aluminum profile on the west facade is crossed at the corner of the road to form the strong visual conflict, which enhances the image tension. On the south side, the whole facade along the street is designed with fluctuating steel aluminum glass and other components forming a strong and dynamic material property. The inte-

rior buildings of the enclosed site create more warm and humane features with lower cost red brick veneer, which is contrasted and reconciled with metal. The chimney of the original induced draft fan room of Building No. 2 is preserved, and the facade with distinct rationalism characteristic is formed by the continuous design of the gable wall. The top of the sunken food plaza is covered with a continuous steel structure with irregular conical roof over the plain weathering steel plate surface. It continues the aesthetic thought of combining contemporary and simple, fully demonstrating the space vitality and richness of the inner courtyard full of leaps and bounds.

This fusion of contemporary and classical spirit conveys the interesting relationship between innovation and memory and humanities in the renovation and transformation of existing buildings. This multiple contrast and reconciliation, which is based on the characteristics of internal and external space, as well as on the space characteristics of formal languages and materials, reveals that under the present social matrix, especially in the new media industrial park, "communication and interaction", as a kind of behavior and space medium, has become the catalyst of "event".

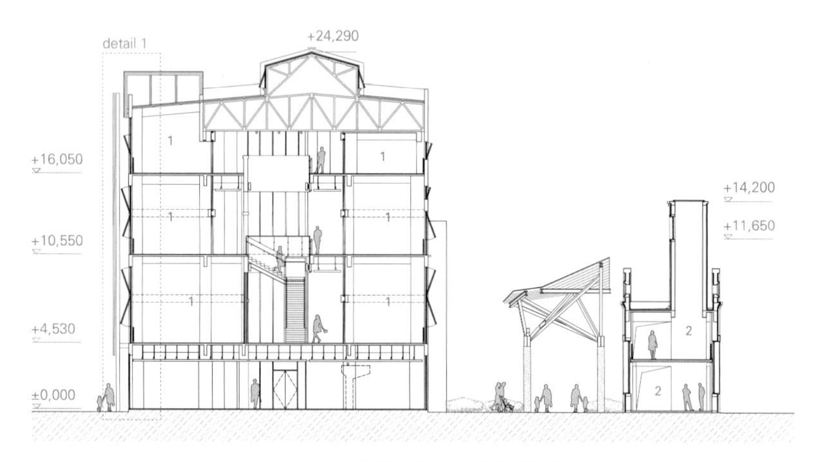

1. 艺术品交易中心 2. 艺术展览画廊
1. artwork trading center 2. artwork exhibition gallery
A–A' 剖面图 section A-A'

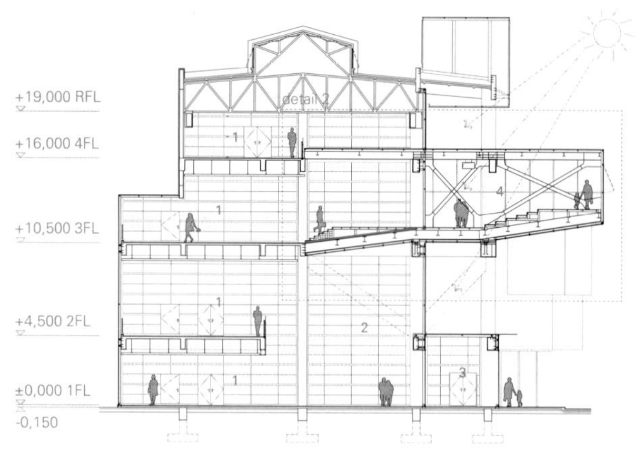

1. 门厅 2. 入口大厅 3. 入口 4. 观景空间
1. lobby 2. entrance hall 3. entrance 4. viewing space
B–B' 剖面图 section B-B'

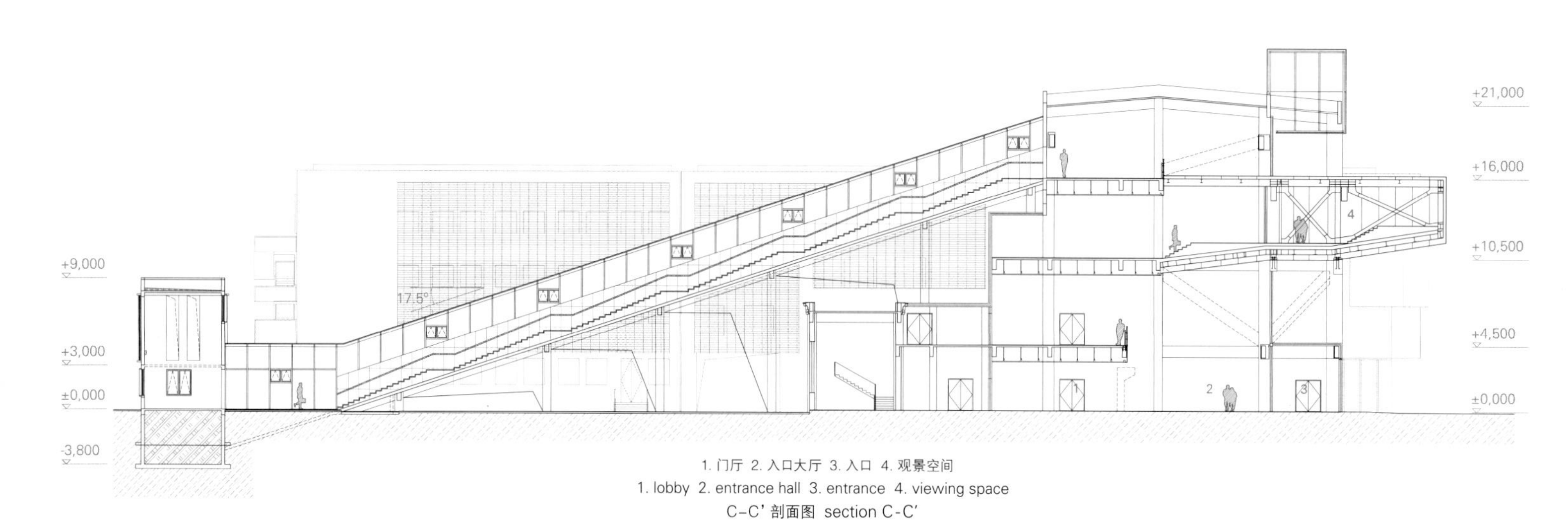

1. 门厅 2. 入口大厅 3. 入口 4. 观景空间
1. lobby 2. entrance hall 3. entrance 4. viewing space
C–C' 剖面图 section C-C'

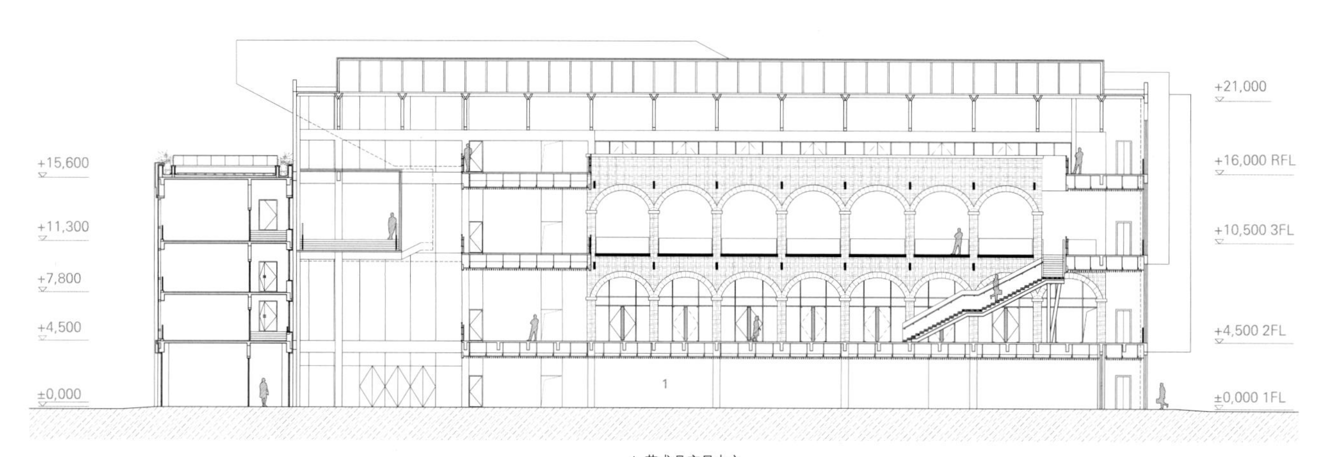

1. 艺术品交易中心
1. artwork trading center
D–D' 剖面图 section D-D'

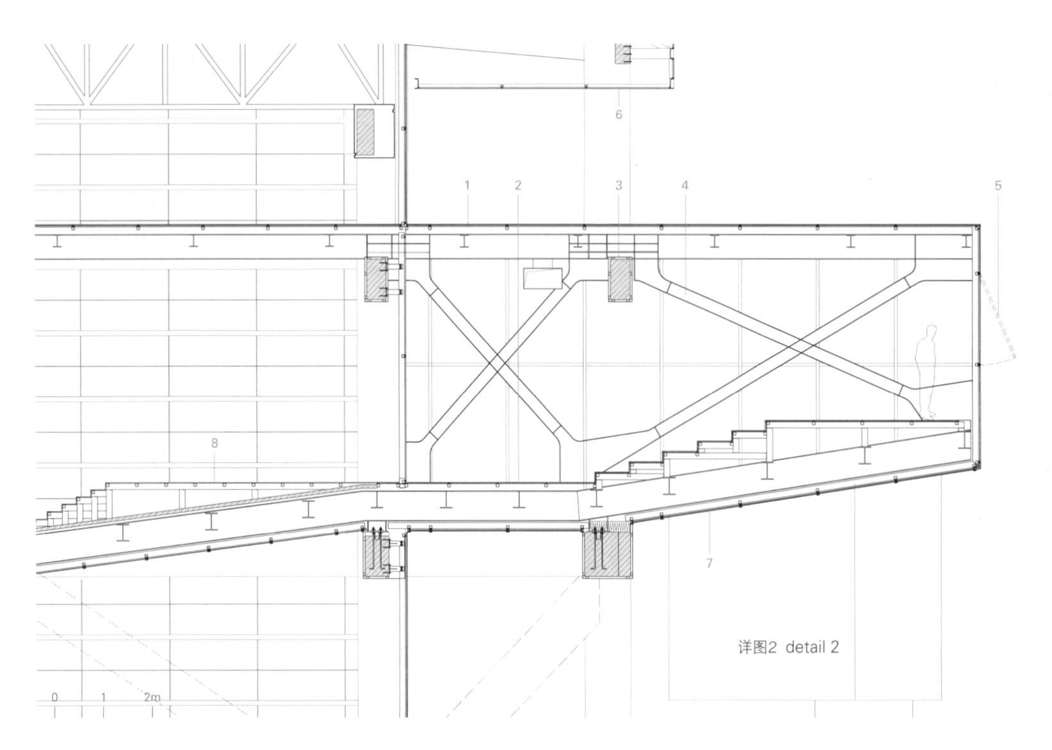

1. 6+12A+6+1.14PVB+6 low-E toughened hollow glass-filled glass frame main keel square tube, 120×60×3, fluorocarbon frame secondary keel steel square tube of the surface, 60×60×3, fluorocarbon I-shaped secondary steel beam I-shaped main steel beam of the surface
2. tempering glass laminated glass TP8+1.14PVB (light tight film), +TP8 frame main keel steel square pipe, 120×60×3, fluorocarbon frame secondary keel steel square tube of the surface, 60×60×3, fluorocarbon I-shaped secondary steel beam I-shaped main steel beam of the surface
3. thickness 3 of dark gray aluminum veneer, 60×40×3 square steel keel
4. I-shaped skew braced steel beam
5. manual overhead ventilation window, 6+12A+6 low-E tempering glass laminated glass
6. thickness 3 of dark gray aluminum veneer, 60×40×3 square steel keel, 20#B box iron
7. 6+1.14PVB+6+12A+6 low-E tempering glass laminated glass, 120×60×4 square steel keel, surface covering fluorocarbon paint, 120×60×4 square steel keel, surface covering fluorocarbon I-shaped main steel beam
8. consolidated composite floor of the thick and solid wood, thickness 5 of riffled plate, 60×60×3 square steel secondary keel, surface covering fluorocarbon paint, 120×6×3 square steel secondary keel, surface covering fluorocarbon paint, frame support square steel tube, 120×60×3 surface covering fluorocarbon paint

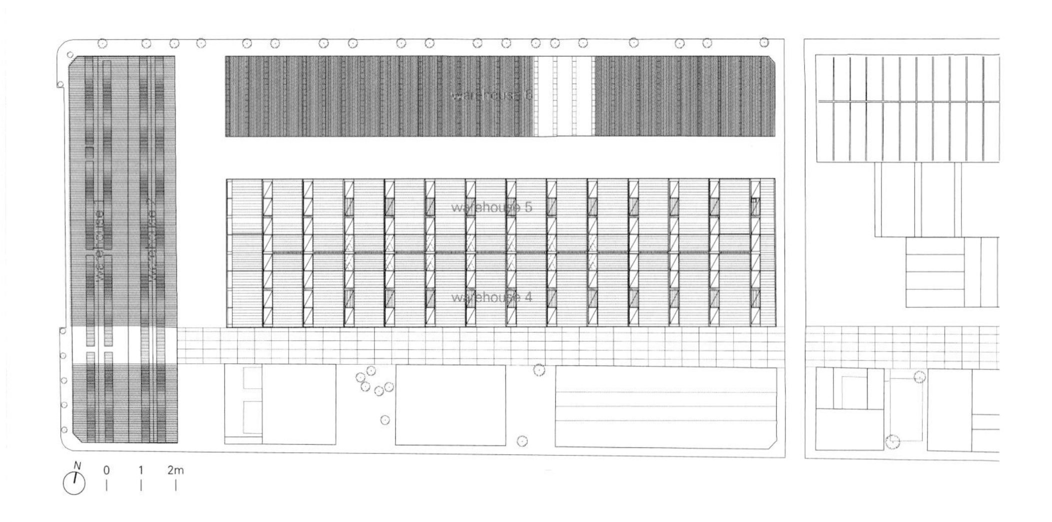

阿尔斯通仓库修复项目
Alstom Warehouse Renovation

Franklin Azzi Architecture

获批重建的阿尔斯通仓库项目位于法国南特岛，该岛正在经历"艺术与历史之城"的创新产业转型，而该改建项目则是其迈向城市复兴的新举措。暂居南特岛已有十年之久的阿尔斯通仓库注定要转型为新的创意艺术区，成为一个文化、科技和经济的交汇点。这片多功能区域占地26 000m²，包括三座建筑——南特圣纳泽尔高等美术学院（ENSBAN）（4号仓库和5号仓库），南特大学（6号仓库），一个容纳创意创业公司、商业加速器、艺术工作室、展览厅和餐饮设施的办公楼（1号仓库和2号仓库）。富兰克林·阿兹建筑设计公司肩负着双重使命：对整个仓库区进行城市规划，以及设计建筑高等美术学院。建筑师拒绝一切哗众取宠、矫揉造作的设计，在设计中不但注重实际效果，而且还注重释放建筑空间，为学生和附近居民提供具有极大灵活性的区域来展示自己的创意和想法。

为了实现一定程度的城市开放性并与周围的城市结构相连接，该项目必须拆除一部分现有建筑，因此使场地恢复了人情味和步行尺度。然而，建筑设计团队珍视这块工业荒地的内在特殊性及其独特的建筑遗产，旨在基于原址记忆对其进行重建。因此，建筑师需要靠想象力来对旧建筑进行重新设计——削减现有的三个体量，打通现有的路径和中间区域，并建造新的建筑。为了保存现有的仓库，建筑师在钢框架体量内部建造了新建筑，同时保留了原始棚屋的屋顶形状。建筑内部体量采用了"俄罗斯套娃"的形式，彼此嵌套，围绕中央内部街道排列。建筑师建造了艺术广场、两条新的步行街和多条通道，通过这三样市政工程开辟出了大型的开放式公共空间，为居于此地的人们提供了会面和交流的平台。滨海艺术中心"集会"位于场地西侧，带顶棚的入口大厅具有防水设计，但却是通风通气的，因此在气候温暖的日子里，露台咖啡馆也会开放。作为创新区的新中心，"集会"艺术中心真正标志着古老的阿尔斯通仓库"堡垒"的开放点。

改建而成的南特高等美术学院有一个专用区域，称为"会议厅"，该区域包括入口大厅、展览厅、图书馆和礼堂。该区域全年向公众开放，人们可以直接从"集会"滨海艺术中心进入该会议厅。在将仓库改造成高等美术学院的过程中，建筑师主要关注如下几点：限制拆除，仅保留钢制框架，引入自然光线，腾出空间以容纳相对密集的居住人口，鼓励提高空间利用率。学校建筑群被划分为两个紧凑的线性箱形体量，它们嵌入各自的棚屋结构，内部街道两侧的众多人行道连接着各个建筑，成为工作室空间的自然延伸。建筑师在实现"俄罗斯套娃"的设计策略时，遵循了建筑结构自我支撑的原则，使新建结构与原有的结构相脱离。为了获得预期的能量和热性能，建筑师使用了两种分离开来的建筑表皮——第一种是聚碳酸酯，第二种是热壳。这种建筑设计方法极为有吸引力，保留下来的工业钢框架上覆盖着聚碳酸酯，就像一把伞，能够根据需要为建筑物提供不同质量的光照。研究实验场地占地4000m²，其内有自由空间、艺术图书馆、创意出版公共中心、青年中心、当代艺术画廊和杂货店。建筑师将工作室设计为开放空间，内部一览无余。还有一些空间尚未定义功能，学生可以自由使用，同时也可留作日后再建。这个雄心勃勃的修复重建项目还旨在促进学院学生之间的会面和集会，在碰撞和交流中激发出创造力。

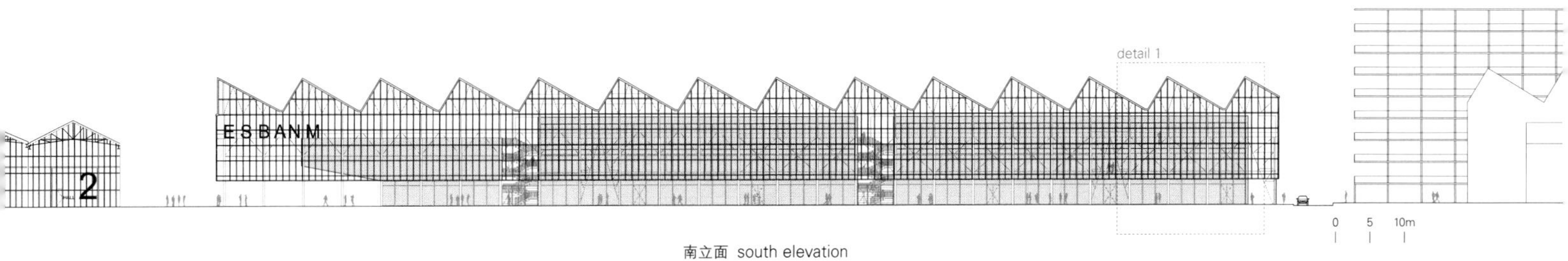

南立面 south elevation

Located in Île de Nantes, the island undergoing an innovative industrial transformation in the "City of Art and History", the Alstom Warehouses requalification project marks a new step towards urban rehabilitation. After 10 years of temporary occupation, the old warehouses were destined to be restructured and converted into the new Creative Art District as the crossroad of culture, technology, and economy. This multipurpose district of 26,000m² comprises three buildings – Nantes Saint-Nazaire Higher School of Fine Arts (ENSBAN) (Warehouses 4 & 5), the University of Nantes (Warehouse 6), an office property complex for creative startups and a business accelerator, artists' workshops, exhibition space, and a catering facility (Warehouses 1 & 2). Franklin Azzi Architecture's mission was twofold: urban planning of the whole warehouse area and architectural design of the Higher School of Fine Arts. Committed to reject-

ing all attempts of architectural gesticulation, this solution not only favors an economy of means but also focuses on freeing spaces that will become expression areas with great flexibility for the students and the neighborhood life.
To establish a degree of urban openness and connect to the surrounding urban fabric, some existing volumes had to be demolished so that the site regains a human dimension and a pedestrian scale. However, the intention was to build on a site's memory, to value the intrinsic exceptional nature of this industrial wasteland and its unique architectural heritage. Thus, it called for an imagination to redesign the site – cutting the three existing volumes, opening up the existing paths and in-between areas, and creating new ones. In order to preserve the existing warehouses, the new buildings are constructed within the volumes of the steel frames and the roofing retains the original sheds' shape. The

项目名称：Alstom Warehouse Renovation / 地点：Mail du Front Populaire – Ile de Nantes - 44200 Nantes, France / 建筑师：Franklin Azzi Architecture / 代表建筑师：Franklin Azzi Architecture / 执行建筑师：ACS / 建筑经济：12ECO / 环境工程：Tribu Ingénierie
声学工程：Lamoureux Acoustique / 立面工程：T.e.s.s. / 总体工程：Setec Bâtiment / 景观设计师：Bureau Bas Smets
技术检验员：Dekra / 客户：Nantes Metropolis, Société d'Aménagement de la Métropole Ouest Atlantique (SAMOA)
委托类型：Urbanism-reconfiguration and conversion of Alstom Warehouses' industrial site, Architecture-rehabilitation and restructuring operation on Warehouses 4 & 5, construction of the Higher School of Fine Arts / 功能：Higher School of Fine Arts, office property complex, restaurant, artist's workshops, and exhibition space / 用地面积：17,000m² / 建筑面积：9,827m² (Warehouses 4 & 5) / 造价：27M € pre-tax value / 竞赛时间：2010 / 施工时间：2014—2017 / 竣工时间：2017.6 / 摄影师：©Luc Boegly (courtesy of the architect)

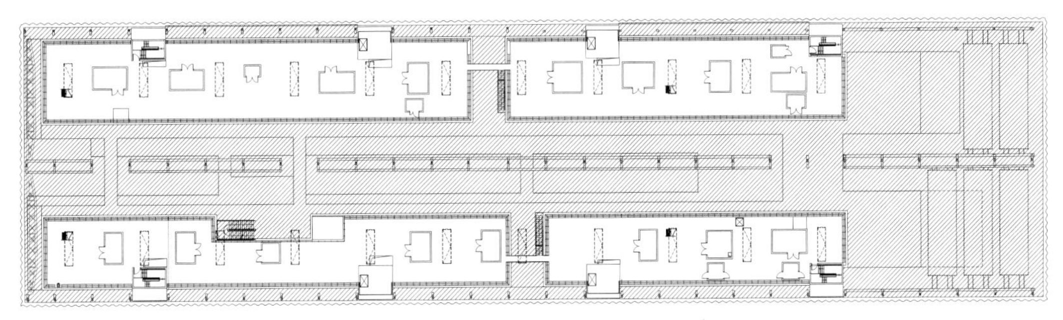

四层 third floor

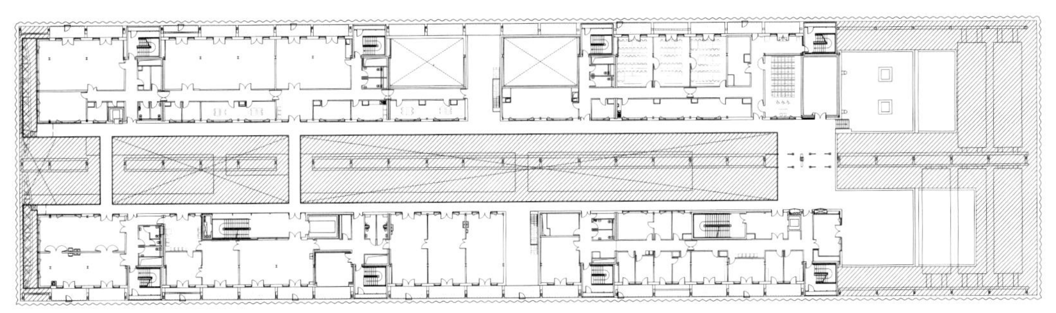

三层 second floor

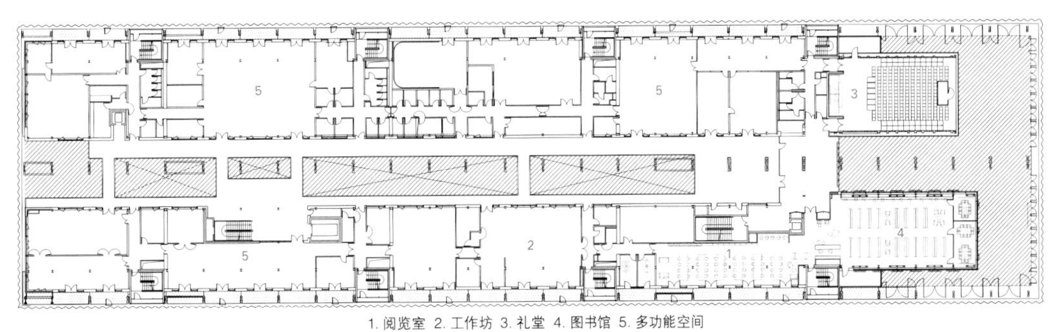

1. 阅览室 2. 工作坊 3. 礼堂 4. 图书馆 5. 多功能空间
1. reading room 2. workshops 3. auditorium 4. library 5. multifunctions space
二层 first floor

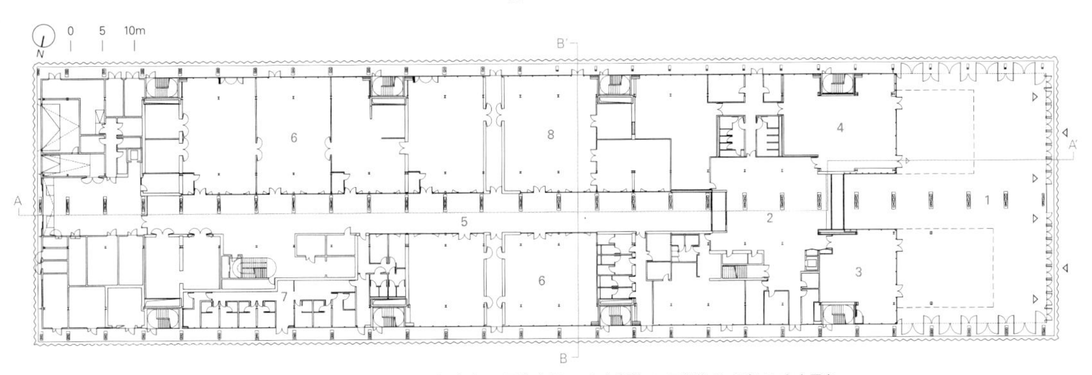

1. 中转广场 2. 入口大厅 3. 阅览室 4. 展览空间 5. 室内街道 6. 工作坊 7. 公寓 8. 自由平台
1. convered concourse 2. entrance hall 3. reading room 4. exhibition space 5. indoor street 6. workshops 7. apartments 8. free platform
一层 ground floor

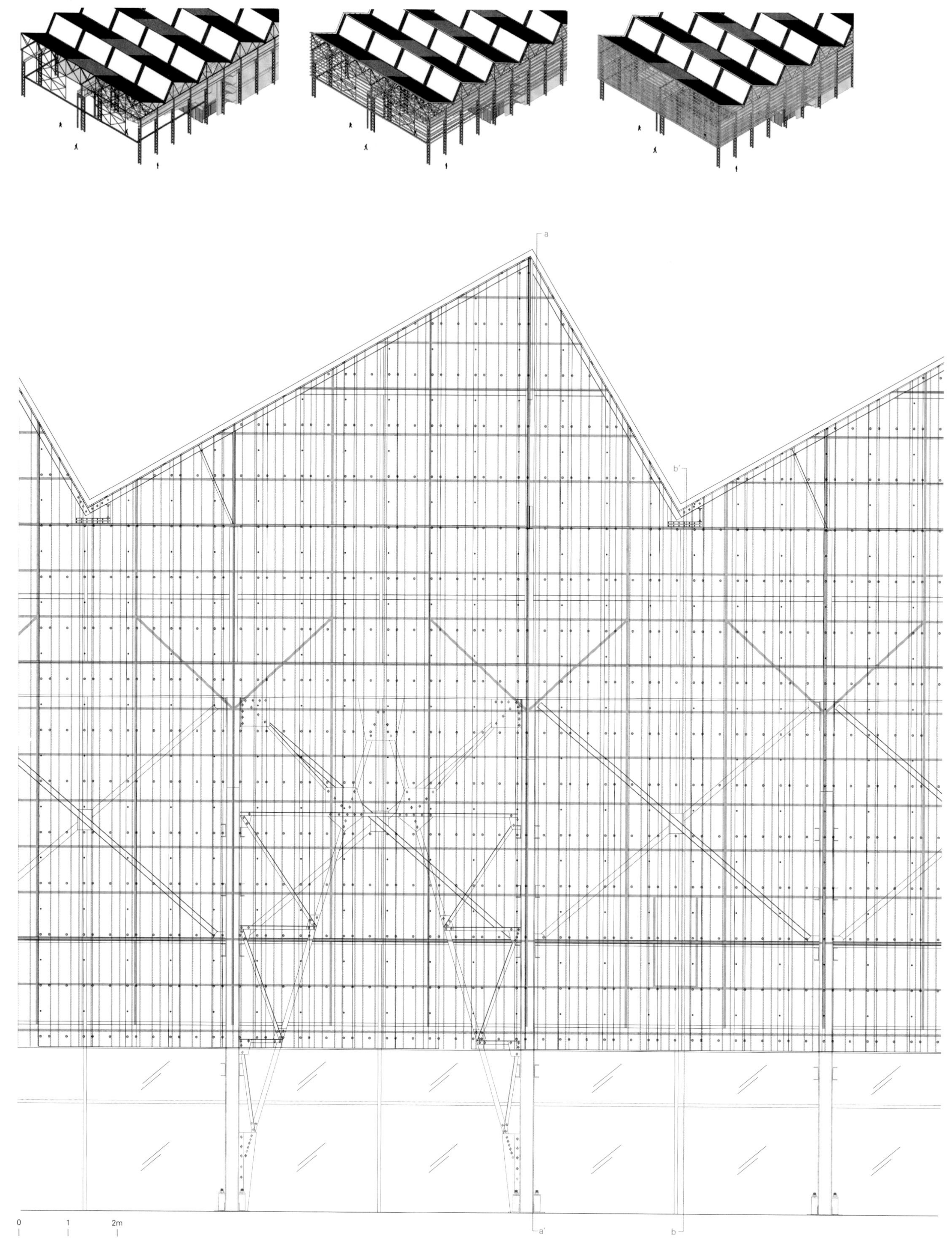

详图1 detail 1

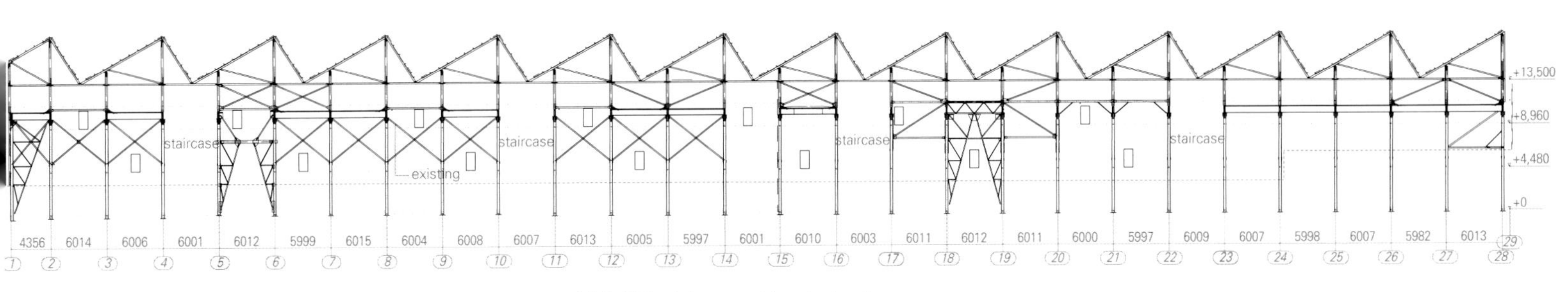

立面加筋图 reinforcement facade elevation

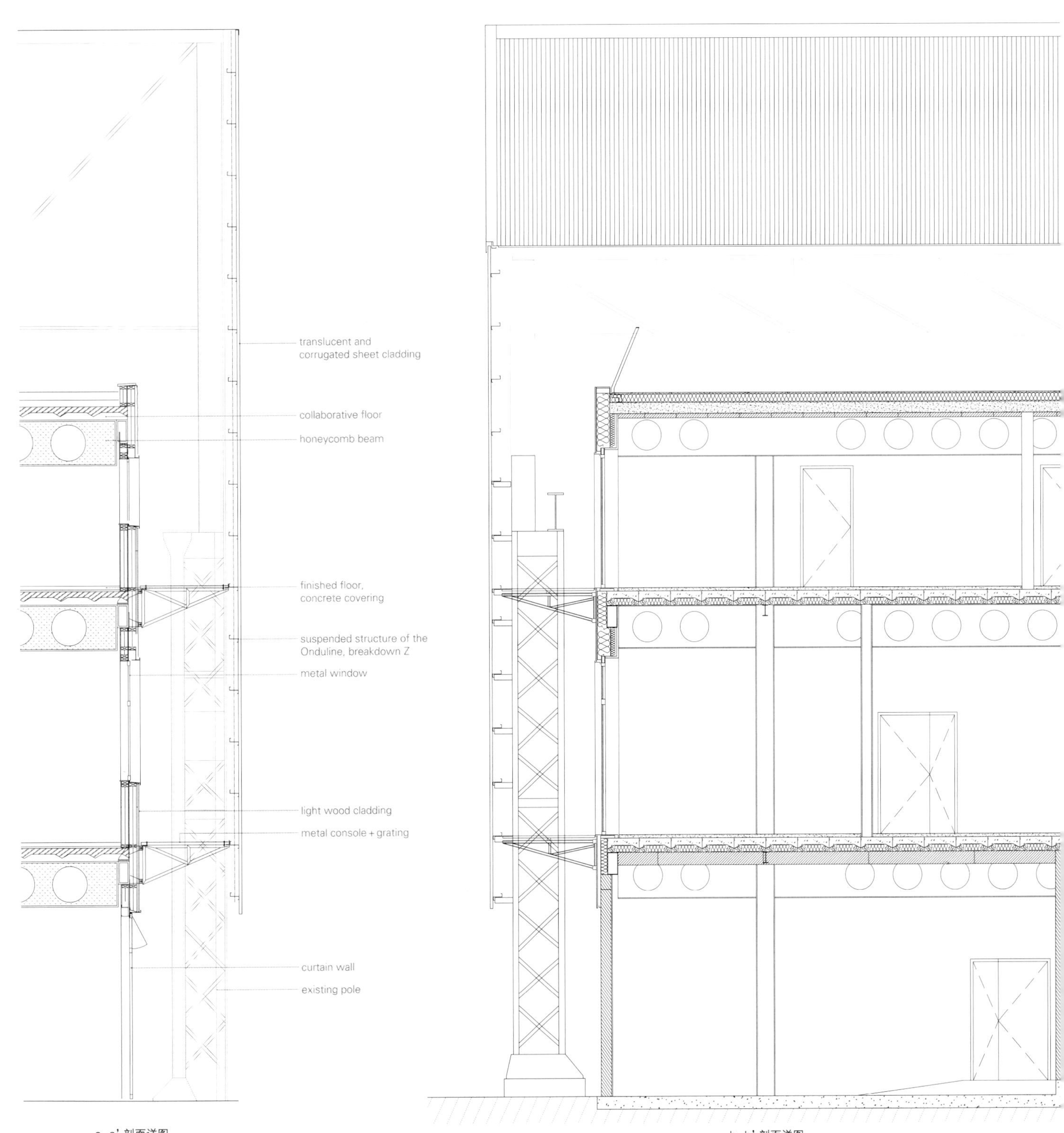

a–a' 剖面详图
detail a-a'

b–b' 剖面详图
detail b-b'

inner volumes take the form of "Russian dolls", nested within each other, arranged around a central interior street. The threefold urban work on the arts esplanade, the two new pedestrian streets, and multiple passageways created large open public spaces where the various dwellers can meet and exchange. On the west side esplanade – the "Agora", the covered entrance hall will be waterproof but not air-proof so that the café terrace can spread out during warm days. As the new centrality of the Creation district, it truly marks the opening point of the old Alstom Warehouses "fortress".

A dedicated part of the school called the "Meeting hall" – entrance hall, exhibition space, library, and auditoriums – is year-round accessible for the general public directly from the "Agora" esplanade. Transforming the warehouses into the Higher School of Fine Arts, the key was limiting demolitions, keeping the steel frames only, bringing natural light, making space for relative density, and encouraging appropriation of spaces. The school premises are divided into two compact linear box volumes inserted in each shed, linked by many walkways on both sides of the inner street that became natural spatial extensions of the workshops. Regarding the strategy for the realisation of the "Russian dolls", a principle of self-supporting structures was executed, so that they are uncoupled from the existing one. To obtain the expected energy and thermal performances, two types of skins are also perfectly uncoupled – the first one is in polycarbonate and the second one is a thermal shell. This approach is architecturally attractive since the preserved industrial steel frames covered with polycarbonate act as an umbrella that gives various qualities of light depending on the needs. The 4,000m² facilities for research and experimentation will be composed of workshops and free-spaces, an art library, a public hub for creative publishing and a youth center, a contemporary art gallery, and a solidarity grocery. The workshops are designed as open spaces so that everyone can see what's going on inside. Some undefined spaces invite students to appropriate the place, allowing a long-term program requalification as well. This ambitious rehabilitation and restructuring project also seeks to facilitate encounters and gatherings among the students in the faculty, that are seen as a vital force for creativity.

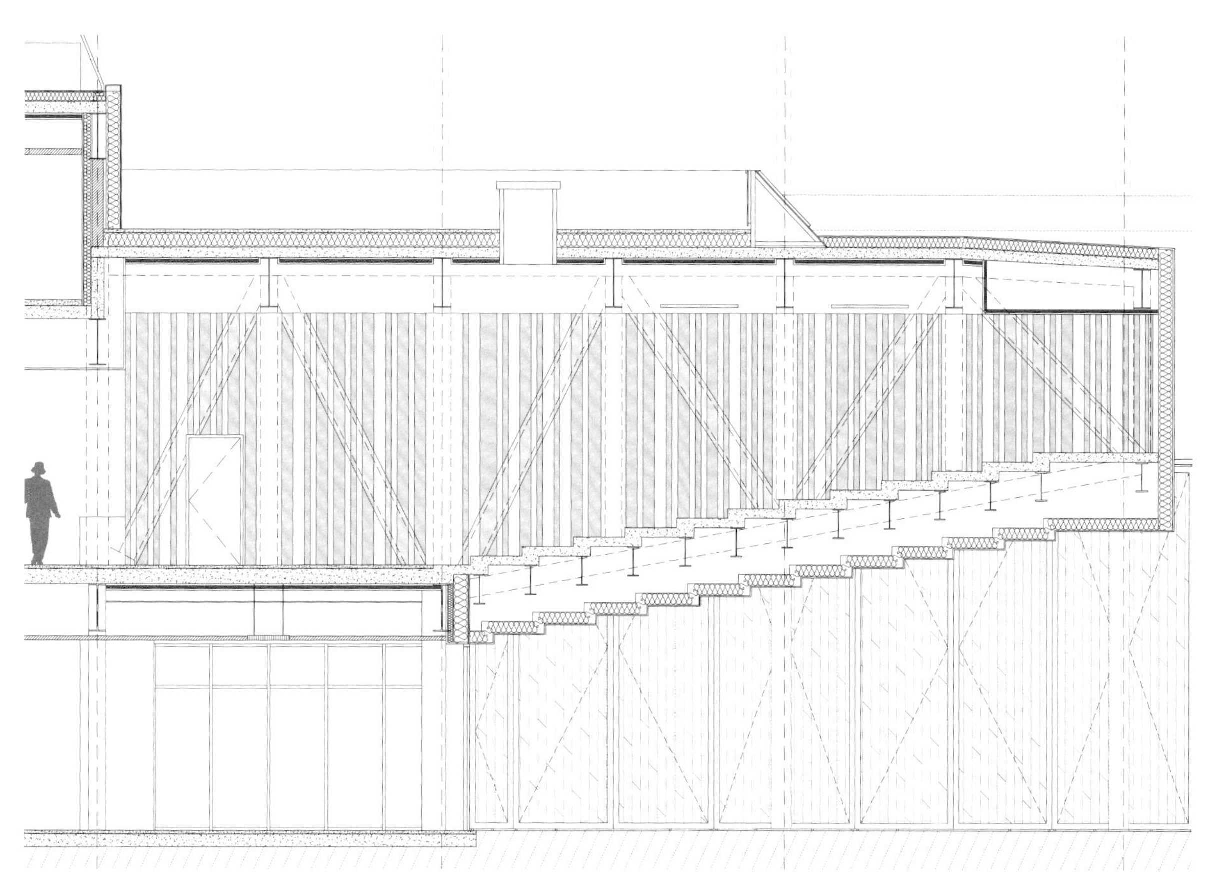

详图2 detail 2

A–A' 剖面图 section A-A'

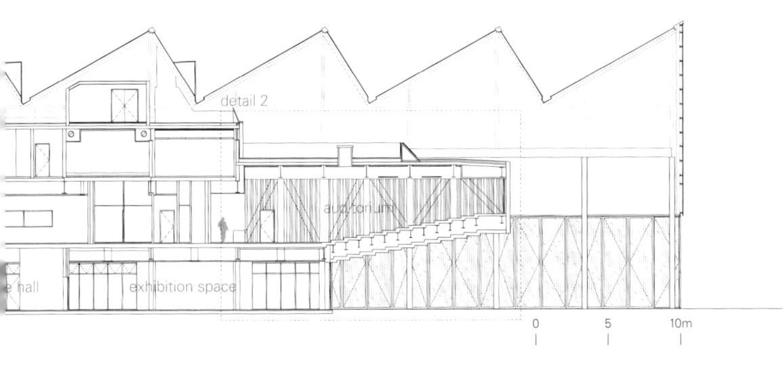

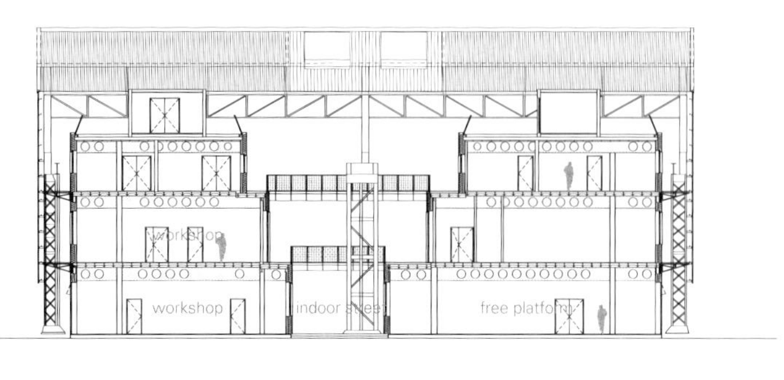

B-B' 剖面图 section B-B'

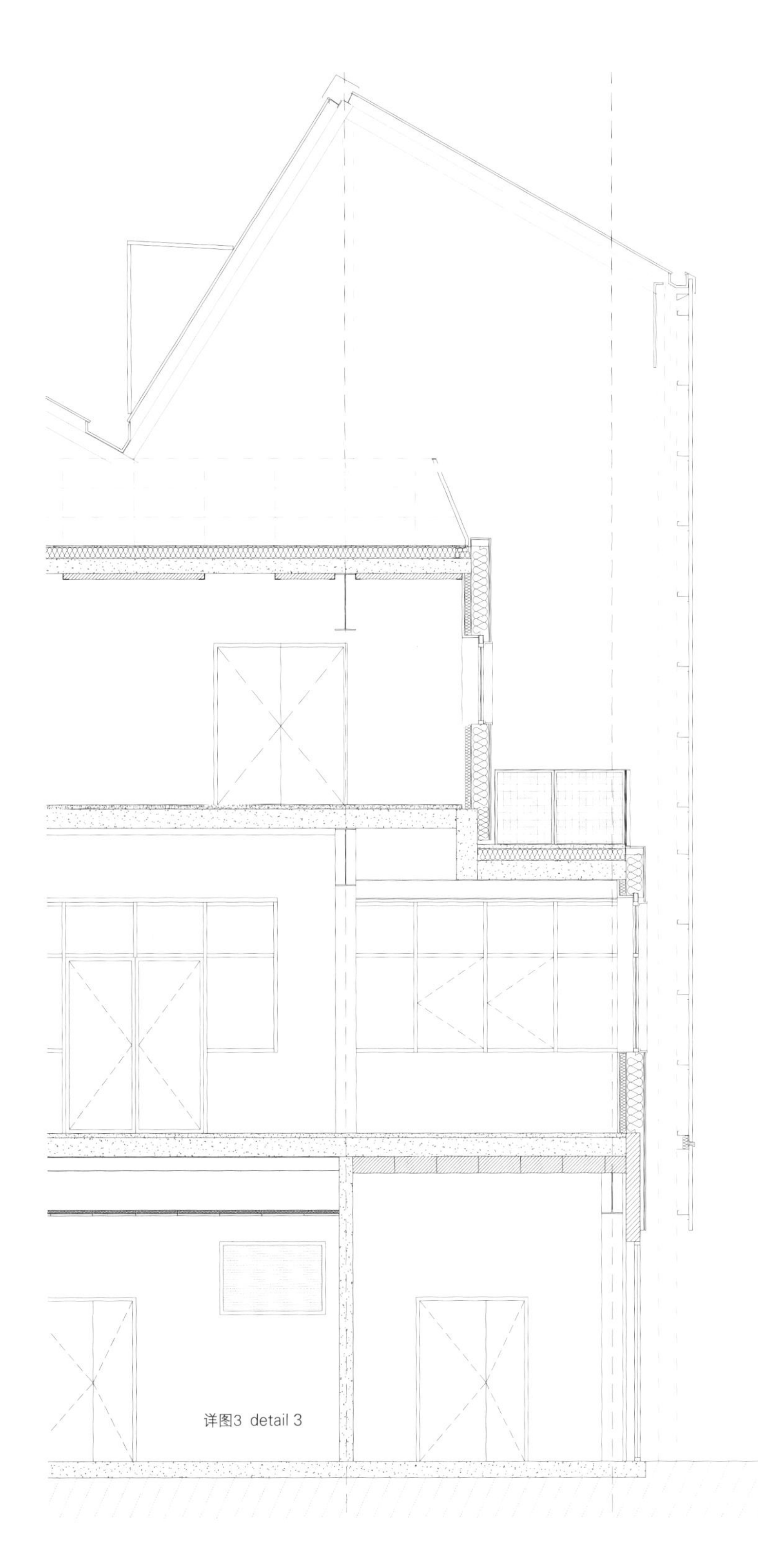

详图3 detail 3

维堡街头圣地
Streetmekka in Viborg

EFFEKT

EFFEKT建筑事务所将一座废弃的风车工厂改造成了街头圣地，使其重新焕发了生机。街头圣地是一个新的文化目的地，配备有跑酷、滑冰、攀岩、篮球、跑步等种类丰富的自发运动设施，并且提供了一系列用于音乐制作和打碟的定制工作室、动画工作室、制造实验室、各种艺术家工作室以及木工和金属工作室。

社交空间和非正式会议区域分布在整座建筑内，并在主要功能区域之间战略性地交织在一起，这一设计理念是通过让参与者接近活动场所来降低参与门槛。

最初的风车工厂是20世纪60年代后期及70年代大规模生产的仓库或工厂的典型建筑。在当时的西方世界，几乎每个郊区工业区都有这样的建筑。人们认为这些由预制混凝土板或波纹钢建成的工业遗物几乎没有什么历史、文化和建筑价值。

EFFKT并不想将遗留下来的建筑拆除，而是想探索如何在非常有限的预算内，以定性的方式对这种被忽视的建筑类型进行重新规划和利用。不管这些盒子形状的建筑外表看起来多么乏味和灰暗，它们的内部空间通常都很宽敞，比例几乎与大教堂一样，结构重复、排列整齐。建筑设计团队只拆除了建筑两端的墙壁，将所有的行政管理功能区域和车间空间都安置在现有建筑的一侧，而滑冰区域则放置在另一侧，以保持原建筑内部制造空间的完整性。建筑的原始结构被保留了下来，成了新建筑的家具元素，原建筑材料也被重新利用，因此翻新成本非常低，最终成本约为新建一座传统运动馆的三分之一。

基于室内街景的概念，建筑师将内向型的工业建筑及其宽敞的中央工厂空间变成了一种新型室内空间：一个对外开放的带顶街景。建筑师基于街景概念，根据特定需求进行相关的建筑功能设计，例如，空间质量、日光、物质性和温度区域。

半透明的功能性聚碳酸酯外壳将新的建筑体量包裹起来，使建筑看起来既轻盈又亲切。同时，新建筑如同一张巨大的画布，当地视觉艺术家可以在建筑墙面上展示他们的艺术作品。这些视觉艺术作品也将新建筑与周围的工业设施清晰地区分开来。

建筑的室内地面自然延伸到外部，建筑周围的休闲绿地上建立了各种街头运动和文化功能设施。未来还会修建步行街和自行车道，将这些活动场地连接到市中心。

维堡街头圣地的内设项目及外观并不是一成不变的。无论是从短期（由于其充满活力的立面）还是长期（由于未来项目功能的更替）来看，它都将与用户一起不断进化和成长。在同一建筑空间内进行着不同类型的活动，这体现了一种共存的理念，从而产生新的协同效应和社会关系。

随着越来越多的空置工业场地被纳入城市扩建计划，维堡街头圣地的设计方法为废弃建筑的整建打下了很好的基础。新社区可以利用这些旧工业建筑来建立其自身的身份和场所感。

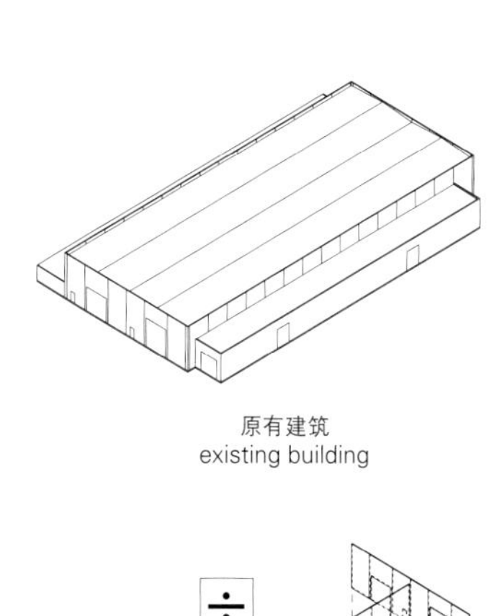

原有建筑
existing building

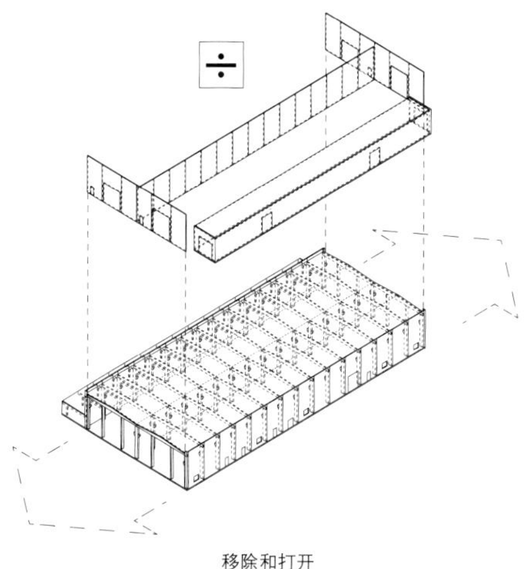

移除和打开
removing and opening up

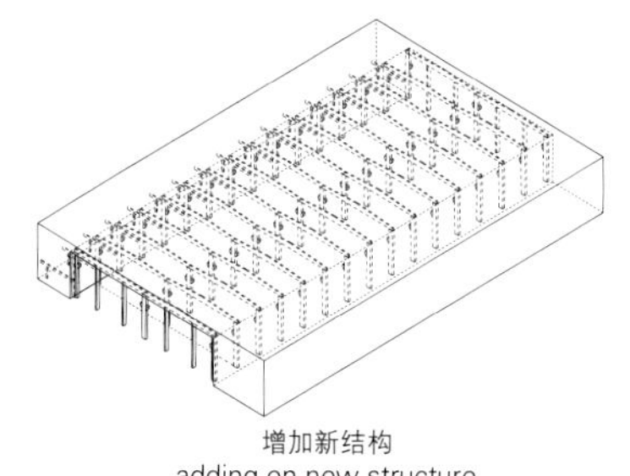

增加新结构
adding on new structure

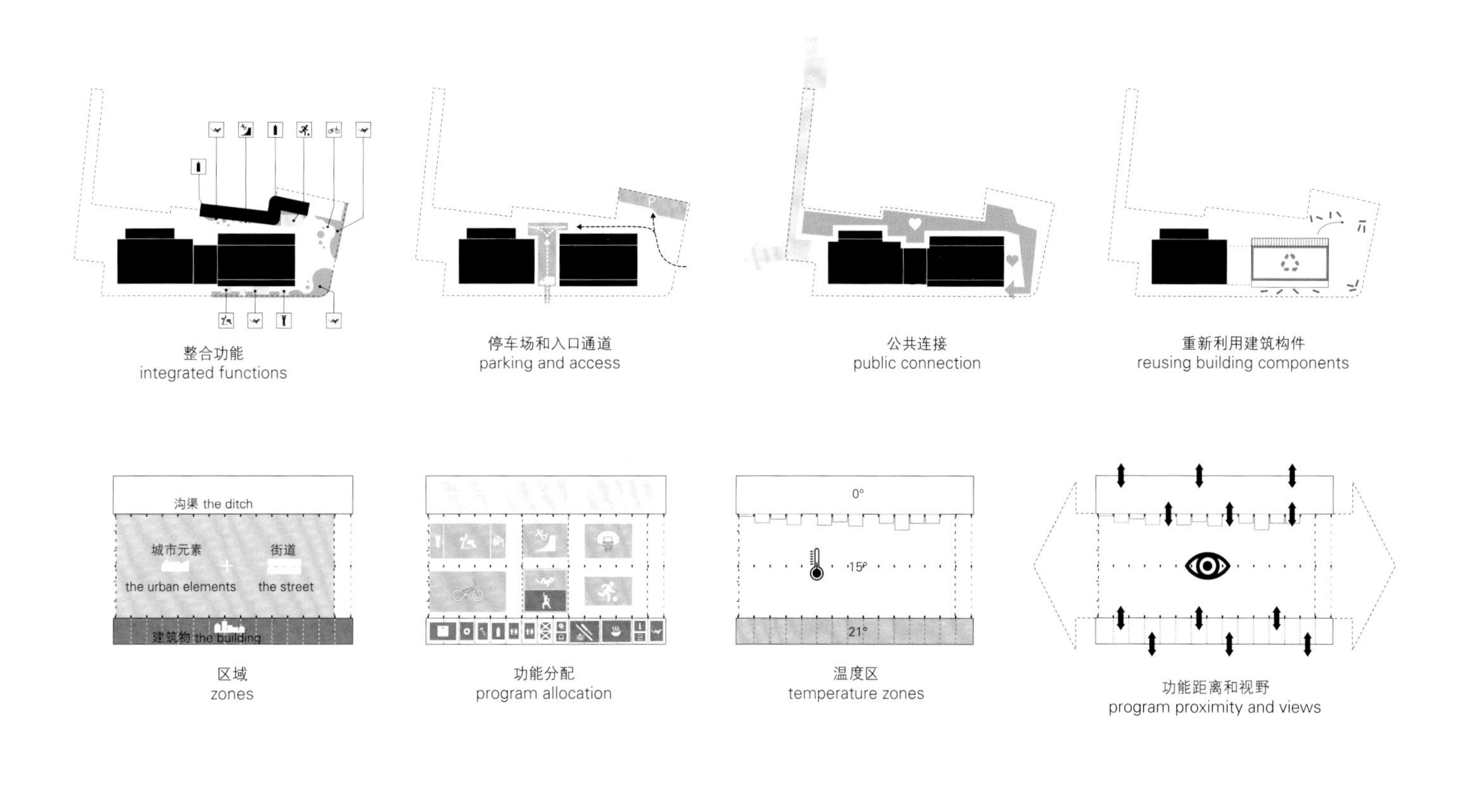

整合功能
integrated functions

停车场和入口通道
parking and access

公共连接
public connection

重新利用建筑构件
reusing building components

区域
zones

功能分配
program allocation

温度区
temperature zones

功能距离和视野
program proximity and views

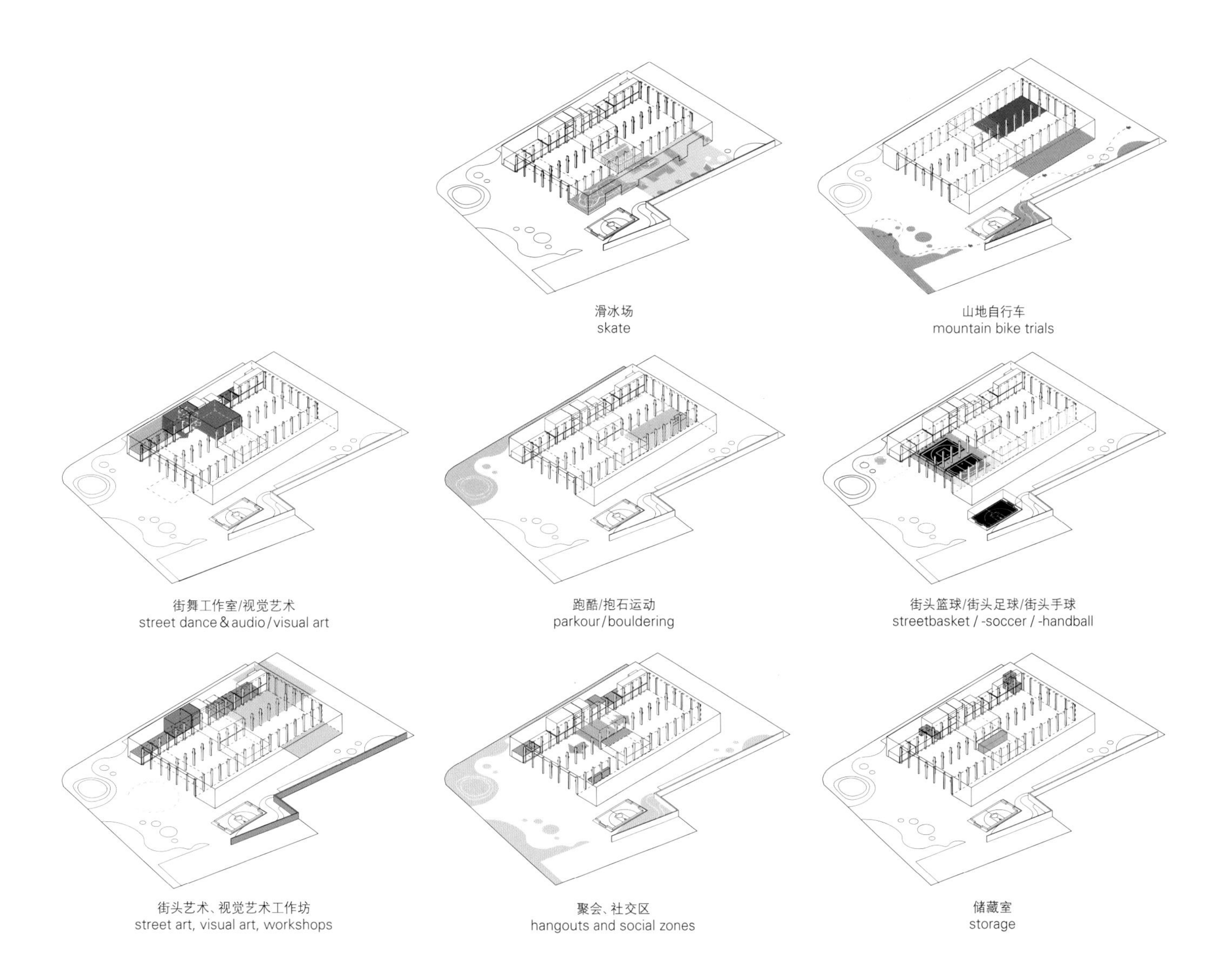

滑冰场
skate

山地自行车
mountain bike trials

街舞工作室/视觉艺术
street dance & audio / visual art

跑酷/抱石运动
parkour / bouldering

街头篮球/街头足球/街头手球
streetbasket / -soccer / -handball

街头艺术、视觉艺术工作坊
street art, visual art, workshops

聚会、社交区
hangouts and social zones

储藏室
storage

Game
WE LOVE
ASPHALT
WE LOVE
ASPHALT

EFFEKT breathed a new life into an abandoned windmill factory – Streetmekka. Streetmekka is a new cultural destination offering a wide variety of facilities for self-organized sports like parkour, skate, bouldering, basketball, trial as well as a series of customized workshop areas for music production, DJ'ing, an animation studio, fabrication lab, and various artist studios and wood and metal workshops.

Social spaces and informal meeting areas are distributed throughout the building and strategically interwoven in-between primary functions based on the notion that proximity to activities lowers the threshold for participation.

The original windmill factory is a typical example of a mass-produced warehouse or factory buildings from the late 1960's and 70's found in almost every suburban industrial zone in the western world. Constructed from prefabricated concrete panels or corrugated steel, these industrial leftovers are perceived as having little or negligible historic, cultural and architectural value.

Instead of demolishing the leftover building, EFFEKT wanted to explore how to re-use and re-program this type of insignificant building typology in a qualitative way, within a very limited budget. However uninteresting and grey the exterior of these boxes appear, they often contain an impressive interior space of magnificent scale and almost cathedral-like proportions based on a repetitive, neatly arranged structural system. After simply removing the walls at both ends of the building, all the administrative functions and workshop spaces are placed on one side of the existing structure and

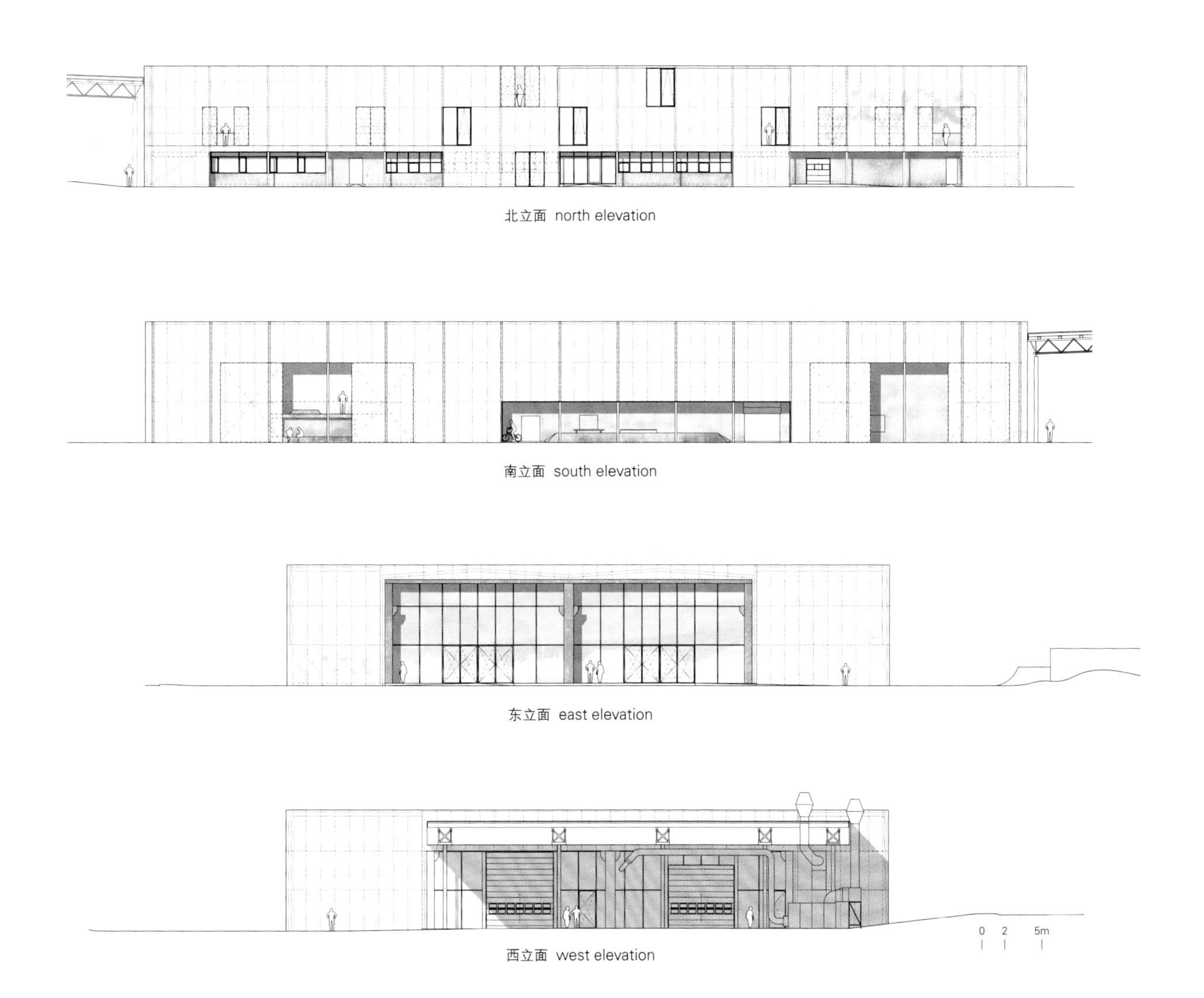

北立面 north elevation

南立面 south elevation

东立面 east elevation

西立面 west elevation

项目名称：Streetmekka in Viborg / 地点：Viborg, Denmark / 建筑师：EFFEKT / 合作方：BOGL, Rambøll, Thomas Andersen A/S / 顾问：Luke Jouppi, Lars Pedersen, Jonathan Linde, Copenhagen Bouldering, Nørlum / 客户：Viborg Kommune and GAME with financial support from Realdania, Lokale- & Anlægsfonden, TrygFonden and NordeaFonden 功能：sports and culture / 用地面积：6,000m^2 / 建筑面积：3,170m^2 + 2,000m^2 landscape / 竞赛时间：2017 / 竣工时间：2018 / 摄影师：©Rasmus Hjortshøj (courtesy of the architect)

DANCE
Game
hummel

SKATE

SK
TE

1. 露台
2. 会议室——动画/视频
3. 会议室——VJ

1. terrace
2. meeting room - animation / visual
3. meeting room - VJ

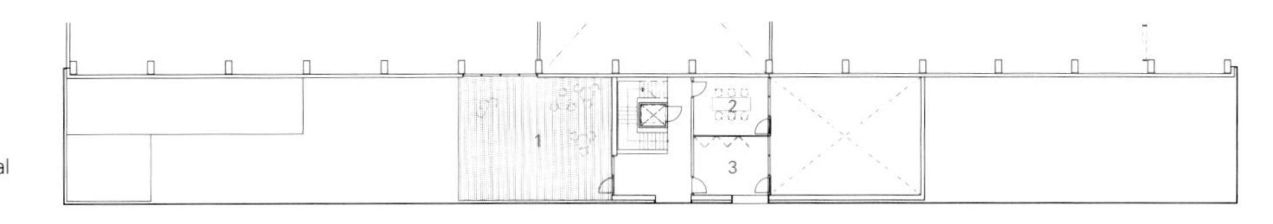

三层 second floor

1. 滑冰场
2. 舞蹈教室
3. 原有的通风道
4. 露台
5. 更衣室
6. DJ/VJ教室
7. 工作坊

1. skate
2. dance room
3. existing ventilation
4. terrace
5. locker room
6. DJ/VJ class
7. workshop

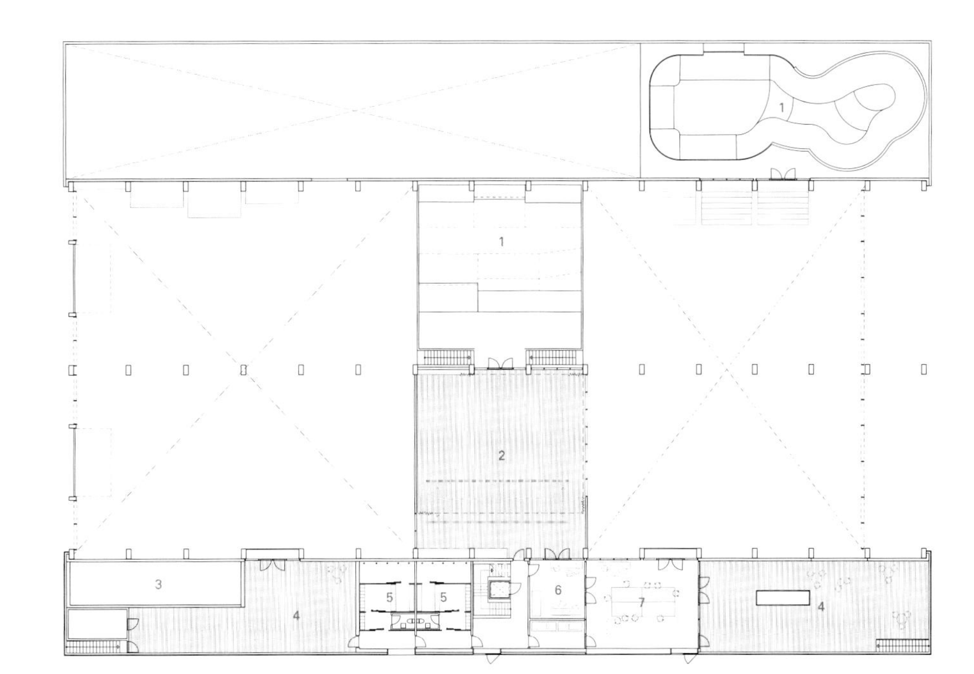

二层 first floor

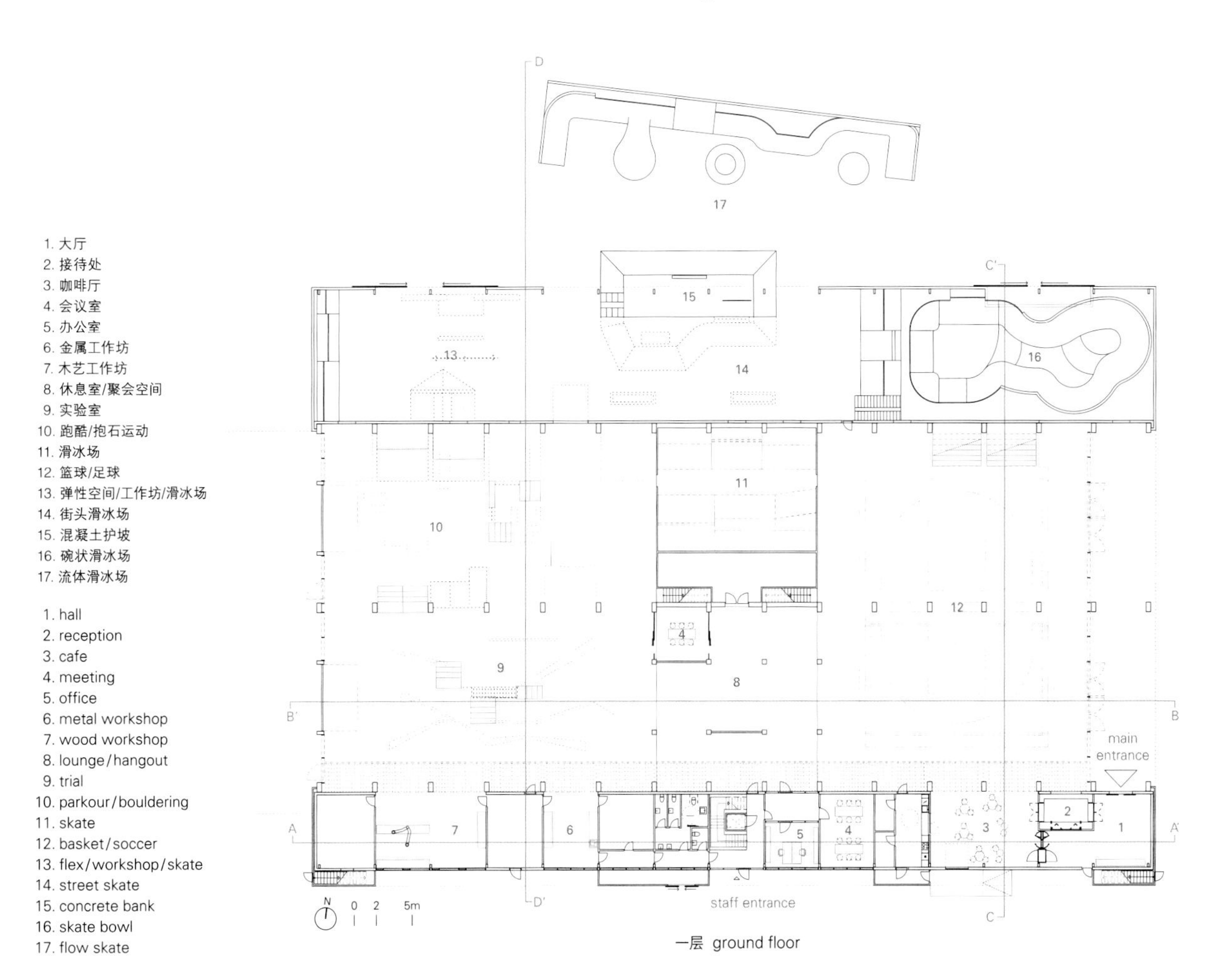

1. 大厅
2. 接待处
3. 咖啡厅
4. 会议室
5. 办公室
6. 金属工作坊
7. 木艺工作坊
8. 休息室/聚会空间
9. 实验室
10. 跑酷/抱石运动
11. 滑冰场
12. 篮球/足球
13. 弹性空间/工作坊/滑冰场
14. 街头滑冰场
15. 混凝土护坡
16. 碗状滑冰场
17. 流体滑冰场

1. hall
2. reception
3. cafe
4. meeting
5. office
6. metal workshop
7. wood workshop
8. lounge / hangout
9. trial
10. parkour / bouldering
11. skate
12. basket / soccer
13. flex / workshop / skate
14. street skate
15. concrete bank
16. skate bowl
17. flow skate

一层 ground floor

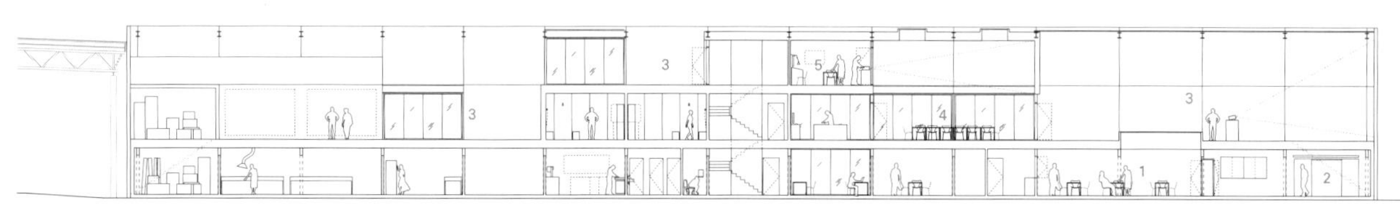

1. 咖啡厅 2. 大厅 3. 露台 4. 工作坊 5. 会议室——VJ
1. cafe 2. hall 3. terrace 4. workshop 5. meeting room-VJ
A–A' 剖面图 section A-A'

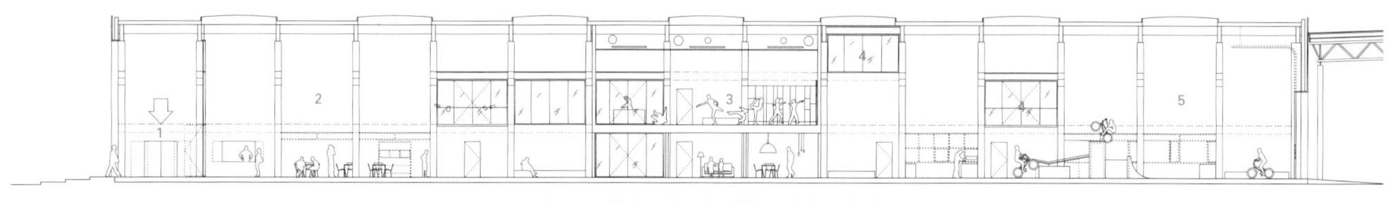

1. 主入口 2. 篮球/足球 3. 舞蹈教室 4. 露台 5. 实验室
1. main entrance 2. basket/soccer 3. dance room 4. terrace 5. trial
B–B' 剖面图 section B-B'

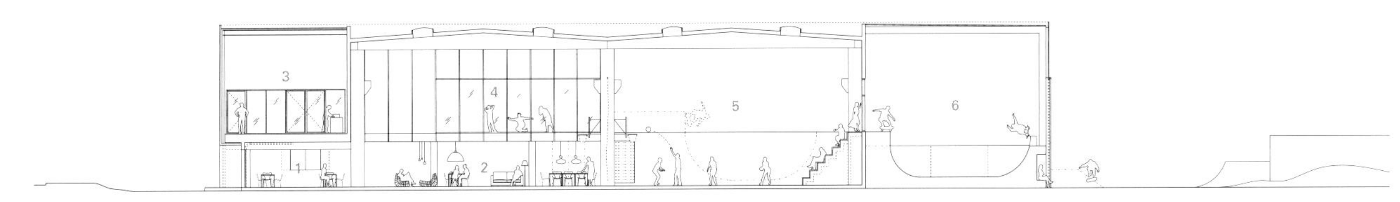

1. 咖啡厅 2. 休息室/聚会空间 3. 露台 4. 舞蹈教室 5. 篮球/足球 6. 碗状滑冰场
1. cafe 2. lounge/hangout 3. terrace 4. dance room 5. basket/soccer 6. skate bowl
C-C' 剖面图 section C-C'

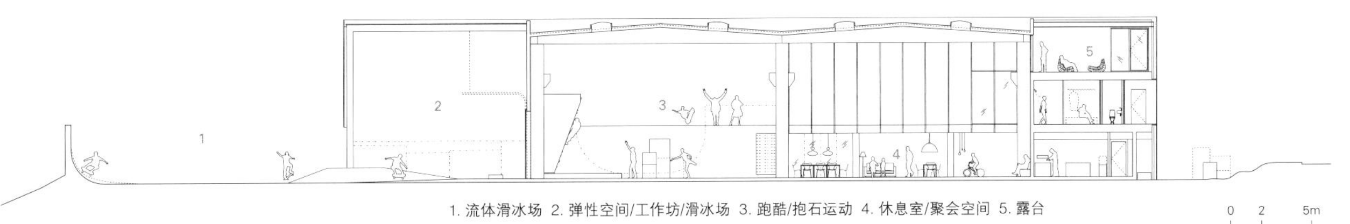

1. 流体滑冰场 2. 弹性空间/工作坊/滑冰场 3. 跑酷/抱石运动 4. 休息室/聚会空间 5. 露台
1. flow skate 2. flex/workshop/skate 3. parkour/bouldering 4. lounge/hangout 5. terrace
D-D' 剖面图 section D-D'

the skate-areas on the other side, leaving the internal former manufacturing space intact. The preservation of the original structure repurposed as furniture elements and the reuse of materials made it possible to carry out the refurbishment at a very low expense, the final cost being approximately one-third of a traditional sports hall.

Based on the concept of an indoor streetscape, the project transformed the introverted industrial building and the impressive central factory space into a new kind of interior space: a covered streetscape open to the outside. The streetscape concept is used to organize the various functions in relation to specific requirements, such as spatial quality, daylight, materiality, and temperature zones.

The new volume is then wrapped with a functional translucent polycarbonate skin, giving the appearance of a light and welcoming building while also serving as a giant canvas for the local visual artists to display and project their art. This also clearly differentiates the building from the surrounding industrial facilities.

The surrounding landscape becomes the natural extension of the indoor surface with various street-sports and cultural functions placed in a recreational string of greenery connecting the site to the downtown area through a future pedestrian and bicycle path.

Streetmekka in Viborg is not static in terms of program and its physical appearance. It will continue to evolve with the users both in short-term – due to the animated facades, and in long-term – due to the future replacement of programs. Housing different types of activity under the same roof enhances the notion of co-existence that breeds new synergies and social relations.

With many vacated industrial sites being incorporated in an urban expansion, this approach paves the way for the revitalization of disregarded buildings left to deteriorate. New neighborhoods can benefit from these industrial heritage markers to build identity and sense of place.

法院
理想的实现和正义的紧张

The Materialisation of the Ideals and Tensions

社会如果没有法治就会陷入大麻烦。正义在开放的空间中运行，但支配它却需要一个密闭的空间。因此，我们有一种建筑类型叫作法院建筑，也称为法庭或司法宫。可以说，人们对法院这一建筑类型的相对忽视，反倒使建筑师在设计法院建筑时免受无政府状态（没有规则）、独裁（有其任意规则）和神权（被认为是超自然力量的规则）的影响，因为在这些建筑中，和谐社会赖以运作的共识规则得以维持。现代社会关于正义的概念起源于古希腊，法院也是如此。

Society would be in big trouble without the rule of law. Justice operates over the open space of territories, but dispensing it requires an enclosure of space. Thus we have the architecture of the courthouse, also known as law courts or palace of justice. It could be argued that the relatively-overlooked courthouse typology defends us from anarchy (which has no rules), dictatorship (with its arbitrary rules) and theocracy (rules credited to supernatural power), because within such buildings, the consensual rules by which harmonious societies can function are maintained. Modern ideas of justice spring from Ancient Greece, and so does the courthouse.
Whether big or small, law courts are projections of civic power into the built environment.

拉里奥加法院_La Rioja Courthouse/Pesquera Ulargui Arquitectos
利摩日法院_Limoges Courthouse/ANMA
科尔多瓦司法大楼_Palace of Justice in Córdoba/Mecanoo Architecten + AYESA
巴黎法院_Paris Courthouse/Renzo Piano Building Workshop
富瓦法院_Palace of Justice in Foix/Philippe Gazeau

理想的实现和正义的紧张_The Materialisation of the Ideals and Tensions of Justice/Herbert Wright

f Justice

无论大小，法院都是公民权力在建筑环境中的投影。传统上，各国都以纪念碑性来标榜自己，蕴含永恒不朽之意，但这与现代政府和社会所期望的透明性截然不同。此外，体现出支持法律的普世价值的建筑，与顺应当地条件和地方风格的建筑有着很大的不同，而当今人们对后者更为敏感和关注。

因此，当代法院建筑在纪念碑性和透明性、普遍性和地方性之间有着有趣的紧张关系。我们以历史为背景，对八座带有古希腊理想印记的当代法院建筑进行了调查，阐释了这些紧张关系是如何演变的。

States traditionally expressed themselves with monumentality, with its implications of permanence, but that's something quite different from transparency, which modern government aspires to and society expects. Furthermore, architecture that expresses the universality of values underpinning law may be quite different from architecture that responds to local conditions and vernacular styles, to which we should be more sensitive today.
Thus, contemporary law court architecture has interesting tensions to address, between monumentality and transparency, between the universal and the local. With a context of historical examples, our survey of eight contemporary courthouses shows how these tensions have evolved, while the echo of Ancient Greek ideals still echoes in the architecture.

理想的实现和正义的紧张

The Materialisation of the Ideals and Tensions of Justice

Herbert Wright

公元前380年，柏拉图在《理想国》中就将对正义的需求与理想国家联系在一起，甚至在更早之前，人们就发现有必要在雅典建立法院。公元前620年，德拉古引入了适用于所有人的成文法律，这就需要进行判决。有时，数千名陪审员聚集在一起，这就需要开放空间，但通常陪审员的数量都在201至501名之间。20世纪70年代，建筑师约翰·特拉沃斯参与了美国古典研究院对雅典卫城下方雅典广场的考古发掘工作，他将发掘出的公元前400年左右遗址的平面图绘制了出来，这些遗址被鉴定为法院（图1）。该遗址有两个相邻的长方形大厅，长度分别约为40m和45m，每个大厅的宽度都是长度的一半，其中一个大厅与一侧开放的带顶柱廊的延伸部分相连，另一个大厅则与两侧围合庭院、边长为49m的方形柱廊相连。如今的法院需要设有办公室、审议室、被告拘留室、档案室、交通循环空间等，而出土的遗址中似乎并没有这些功能区域。然而，这些遗址仍可能是法院类型建筑的根源。从历史上来看，新古典主义风格的大法院往往参考古希腊风格而建。建筑师将门廊入口扩大，成为一个全宽的纪念柱柱廊，列于山墙之下，传达出永恒与智慧之感。罗马人将这种建筑与文艺复兴后的建筑相结合，成为欧洲宏伟建筑的模板。最终，这种建筑成了19世纪美国大法院的一个标志，随着CITES和法律职业的发展壮大，这类法院的数量和规模也不断扩大。这种建筑风格在卡斯·吉尔伯特1935年完成的美国最高法院（图2）中达到了顶峰，该建筑的外型就如同装上了厚重而坚实的翅膀的雅典卫城。

Even before Plato connected the need for justice to the ideal state in his Republic discourses in about 380 BC, there was a need for a place for courts to convene in Athens. In 620 BC, Draco introduced written laws that applied to everyone, which required adjudication. Sometimes, thousands of jurors were assembled, requiring open space, but normally, there were from 201 to 501 jurors. In the 1970s, architect John Travlos, who worked with the archeological excavations by the American School of Classical Studies at the Athenian Agora, below the Acropolis, drew plans of excavated ruins from about 400 BC which were identified as courthouses[figure 1]. There were two adjacent rectangular halls about 40m and 45m long and each half as wide, one joined to a covered colonnade extension open on one side, and elsewhere, a square peristyle colonnade of 39m on each side enclosing a courtyard. Today's courthouses require offices, deliberation rooms, defendants' holding cells, archives, circulation space and more, which seem to be lacked in these unearthed ruins structures. Nevertheless, they are likely to be the roots of the courthouse typology. Historically, the neoclassical styles of great law courts often referenced Ancient Greece. The portico entrance expanded to a full-width monumental colonnade of columns under a pediment, conveying a sense of timelessness and wisdom. The Romans appropriated it and post-Renaissance, which became a template in European grand architecture. Finally it became a hallmark of great nineteenth century American courthouses, which grew in size as cites and the legal profession expanded. This culminated in Cass Gilbert's US Supreme Court[figure 2], finished in 1935, which in form is as if the Acropolis had been given bulky, solid wings.

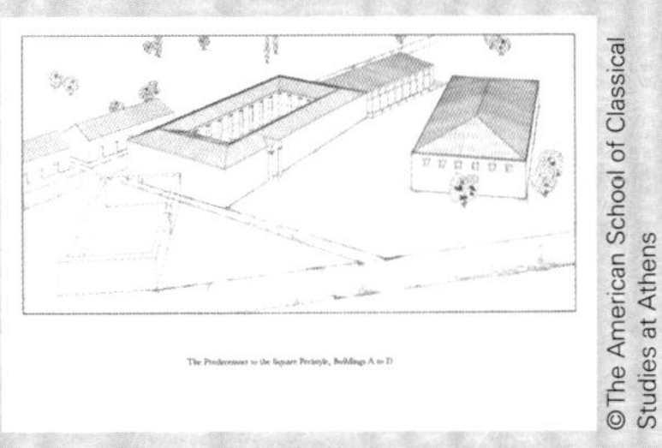

© The American School of Classical Studies at Athens

1. 雅典广场的前身，A楼到D楼
1. The Predecessors to the Square Peristyle, Building A to D

©Joe Ravi

2. 美国最高法院，卡斯·吉尔伯特，华盛顿特区，1935年
2. US Supreme Court by Cass Gilbert, Washington, D.C., 1935

3. 杜瓦尔县法院，KBJ建筑师事务所，佛罗里达州，2010年
3. Duval County Courthouse by KBJ Architects, Florida, 2010

在当代建筑中，这种全宽入口的宏伟建筑形态非常罕见，但在我们的调查中，有一个法院却以一种人们意想不到的形式拥有它。由Pesquera Ulargui Arquitectos建筑师事务所设计的西班牙拉里奥加法院（108页）高四层，每层平行的长条空间看起来和商业园区中的普通办公室隔间几乎无异，阳光通过楼层之间的空隙射入建筑物内部，同时这些空隙也形成了种植有稀疏植被的长露台。有两个特殊的元素连接着这些体块。首先，长方形空间的三面墙体皆为外框墙，墙上每层都种有植被，在建筑周围形成了具有生命的幕墙。这种设计将植被融入了建筑的外壳之中，有利于保护环境和人们的健康，并顺应了世界新兴的建筑趋势。但最引人注目的元素是对军事医院旧址的重新利用，这家砖面墙体的医院贯穿了整个场地的北部。医院大楼已被完全拆除，只留下围墙，形成了一个两层楼高的长条形公共大厅，作为新法院的入口。在该项目中，旧建筑并不具有任何形式的古希腊建筑风格，但却和一座宏伟的希腊式柱廊有着异曲同工之妙，体现出了法院的建筑风格和庄严感。

雅典罗马式柱廊和山墙在后现代主义建筑设计中短暂重现，而KBJ建筑师事务所位于佛罗里达州杰克逊维尔的杜瓦尔县法院（2010年）（图3）就是这样一个建筑范例。但如今的现代主义已经不再参考历史建筑风格，即便是正在修建中的华盛顿最高法院，也将摆脱新古典主义设计的沉闷感和精英主义。

The grand gesture of the full-width entrance extension is rare in contemporary architecture, but one courthouse in our survey has it in an unexpected way. Pesquera Ulargui Arquitectos' La Rioja Courthouse (p.108) in Spain has four long parallel volumes looking like undistinguished offices in a business park, with spaces between which bring light into the buildings, and create long patios planted sparsely with trees. Two extraordinary elements link the blocks. First, three sides of the rectangular space containing the new volumes have an outer frame wall, planted with vegetation on each floor that creates a living screen around them. This places the project in the emerging worldwide trend of buildings where vegetation is integrated into the envelope, with resulting environmental and wellbeing benefits. But the most dramatic element is the re-use of the old brick-faced military hospital which runs the entire length of the site's northern side. This has been entirely stripped back to just its perimeter walls, to become a long two-storey public hall, which serves as the entrance to the new law courts. The old building has nothing of the style of Greek architecture, but it has a similar effect to a grand portico, both in its architectonics and contributing gravitas to the courthouse.

Athenian-Roman porticos and pediments would temporarily re-emerge in post-modernism, and a late echo of that is KBJ Architects' Duval County Courthouse in Jacksonville, Florida (2010) [figure 3]. But modernism had already dispensed with historical reference even as Washington's Supreme Court was being built, and it would go on to liberate the courthouse typology from the stuffiness and elitism that formulaic neoclassical designs had come to embody.

4. 埃弗雷特·麦金利·德克森法院，密斯·凡·德·罗，1964年
4. Everett M. Dirksen U.S. Courthouse by Mies van der Rohe, Chicago, 1964

5. 印度昌迪加尔高等法院，勒·柯布西耶，1952年
5. High Court by Le Corbusier, Chandigarh, 1952

©Sanyam Bahga

6. 利摩日法院，ANMA建筑事务所，法国
6. Limoges Courthouse by ANMA, France

©Sergio Grazia

有两位卓越的现代主义建筑大师在他们的晚年设计了伟大的法院建筑作品。勒·柯布西耶设计的印度昌迪加尔高等法院（1952年）（图5）是他最杰出的作品之一。柯布西耶设计了一座带有刻纹装饰的混凝土长块型建筑，其建筑正立面不与地面垂直，并被入口一分为二。入口为三座色彩明艳的18m高墙，像柱子一样，向上延伸，在法院上方撑起了沉重的、与原建筑分离的第二个屋顶。另一位建筑设计师路德维希·密斯·凡·德·罗是现代主义建筑师中最崇尚极简主义的一位，他设计的一座法院建筑体现出了其晚期美国设计的精髓——幕墙以及垂直伸出的黑钢和黑玻璃矩形结构。美国的埃弗雷特·麦金利·德克森法院（1964年）（图4），也被称为德克森联邦大楼，共有30层，高117m。这座平板状建筑是芝加哥三大联邦广场大楼之一。最初，大楼的上层有15个法庭（现在有更多）以及联邦司法机构的办公室。具有讽刺意味的是，同样是在芝加哥，1968年民主党大会上的抗议活动遭到了猛烈的镇压，暴露出了当时美国司法中的深层次缺陷。

ANMA建筑事务所设计的法国利摩日法院（126页）（图6）的显著特征是其立方体外形，或者更准确地说，是一个几乎成立方体的空隙。利摩日法院是一座看起来很清爽的白色建筑，四层楼的外墙垂直矗立于地面之上，白色悬臂屋顶延伸出来，俯瞰其下平淡无奇的硬地广场。两面空白的白色墙壁凸显了中间的空隙，看起来像一个门廊，被延伸出来的建筑所覆盖，居于建筑群落的一角。进门之后，中间走廊的设计继续体现了令人眼花缭乱的美学，走廊墙面朝着轴向天窗倾斜，体现出设计的不同寻常之处。如雕塑一般的走廊中有着夸

Two of the supreme modernists would design great courthouses in their later life. Le Corbusier's High Court in Chandigarh, India (1952)[figure 5] is one of his most extraordinary works — a long sculptured concrete block with off-vertical facades, divided into two parts by an entrance through three brightly coloured 18m-high walls that, like columns, reach up to a heavy, separated second roof above the whole courthouse that shades it. The most minimal of modernists, Mies van der Rohe, designed a legal building which embodies the essence of his late American designs – the curtain-walled, vertically-extruded orthogonal volume of dark glass and steel. The Everett M. Dirksen U.S. Courthouse (1964)[figure 4], also known as the Dirksen Federal Building, is a 30-storey, 117m-high slab and one of a composition of three Federal Plaza buildings in Chicago, and originally had 15 courtrooms on its upper floors (it now has more) as well as offices of federal justice agencies. Ironically, it was in Chicago in 1968, when protests at the 1968 Democratic Convention were violently suppressed, that deep flaws in American justice at the time were exposed.
A cube, or more precisely, an almost cubical void, is the distinguishing feature of ANMA's Limoges Courthouse (p.126) in France[figure 6]. This is another crisp, white building with four-storey facades which stress verticality, but it is entered over a featureless hard-surfaced square that exactly maps on to an extension of the roof cantilevered out above it. The covered void, emphasised by blank white walls on two sides, is almost like a portico in that the building extends to cover it, but on the corner of the block. The dazzling aesthetic continues inside the central corridor beyond the door, which unusually has walls angled in towards the axial skylight. The dramatic square parasol

7. 纽约时报大楼，RPBW，2007年
7. New York Times Building by RPBW, 2007

8. 碎片大厦，RPBW，伦敦，2012年
8. The Shard by RPBW, London, 2012

9. 巴黎法院，RPBW，巴黎，2017年
9. Paris Courthouse by RPBW, Paris, 2017

10. 正义宫，约瑟夫·波勒特，比利时布鲁塞尔，1884年
10. Palace of Justice by Joseph Poelaert, Brussels, 1884

张的方形阳伞元素，这就减少了对于柯布西耶在昌迪加尔法院设计中的雕塑自由理念的呼应。

伦佐·皮亚诺一直深谙建筑设计中的透明艺术，这一理念体现在他对高层建筑的设计之中，例如，伦敦碎片大厦（2012年）（图8）和纽约时报大楼（2007年）（图7）。结构上的轻盈感和内部充足的光线是他的RPBW建筑事务所的标志。他最近的建筑设计作品——巴黎法院（154页）（图9）高达160m，是欧洲最高的司法大楼。巴黎法院大楼的基座是一座大型L形裙楼，楼顶种满树木，上方堆叠了三个玻璃箱板结构，每个结构都高达九层，用凹形地板垂直隔开。每一个玻璃箱板结构都比其下边的短一截，就像楼梯一样，因此玻璃箱板结构上的两个花园梯田可以顺势与大平台上更为宽阔的屋顶花园梯田相连。这座不同凡响的38层建筑是正义的标志，代表着正义在巴黎天际线上独特的存在。或许只有另外一座法院可以与之相媲美——约瑟夫·波勒特设计的比利时布鲁塞尔正义宫（1884年）（图10），它同样孤独地矗立在一个大都市的上空。巴黎法院是一座超现实主义、折中风格的高密度建筑，运用了大量的古典柱子，穹顶高达104m。但是它的巨大重量和RPBW其他巴黎设计项目的轻盈之间存在了巨大的差异，这表明了人们对于正义的预期已经发生转变，从帝国和独裁转变为民主和透明。

同时，巴黎法院带有另一层含义，关于巴黎这个城市本身。巴黎法院坐落在克里希港，距离环城大道不远，这条公路将市区与拥有大量社会住房的广阔郊区分隔开来。因此，巴黎法院在大巴黎城市议程中扮演了消弭环城大道两侧社会和地理鸿沟的角色。法律界不

element and the sculptural corridor are almost distant, diminished echoes of the sculptural freedom Corbusier exercised in Chandigarh.

Renzo Piano has long mastered the art of transparency in buildings, including high-rise such London's Shard (2012) figure 8 and the New York Times Building (2007) figure 7. A sense of lightness in structure and an abundance of light within are hallmarks of his RPBW practice. The 160m-high Paris Courthouse (p.154) figure 9, Europe's largest judicial building, is the latest example. Above a large L-shaped podium building planted with trees on its roof, three nine-storey glass box slabs are stacked, separated vertically by recessed floors. Each is stepped back at one end from the volume below, like stairs, enabling two garden terraces to cascade down to the podium's wider roof garden terrace. The extraordinary 38-storey structure landmarks justice as a unique presence on the Paris skyline. Perhaps only one other law court has stamped such a strong solitary identity visible across a metropolis – Joseph Poelaert's Palace of Justice (1884) in Brussels, Belgium figure 10. This surrealistically dense and eclectically styled building, with plenty of classical columns, has a dome reaching 104m. But the contrast between its sheer heaviness and the lightness of RPBW's Paris design could not be greater, indicating how the intended projection of justice has changed, from imperial and authoritarian to democratic and transparent.

The Paris Courthouse signifies another intention too, about the city itself. It is situated at Porte de Clichy not far from the Boulevard Périphérique orbital highway that separates the city proper from the extensive suburbs characterised

11. 苹果公司总部，福斯特及合伙人公司，加利福尼亚州库比蒂诺，在建
11. Apple Park by Foster + Partners, Cupertino, California, under construction

12. 富瓦法院，Phillipe Gazeau
12. Palace of Justice in Foix by Philippe Gazeau

愿意从巴黎圣母院附近的旧法院搬到一个乏味的城市边缘地带，但随着克里希地区的复兴，这样的抱怨在巴黎应该越来越少了。

建筑师Phillipe Gazeau设计的富瓦法院坐落在法国南部小城富瓦（172页）（图12）边上一座树木繁茂的山丘之下，远离城市的喧嚣。该建筑最引人注目的元素是一条蜿蜒的斜坡，通向一座悬浮的圆形建筑，与其后边的山丘相连接。这座建筑并不像福斯特及合伙人公司设计的苹果公司总部，也就是位于加利福尼亚州库比蒂诺（在建）（图11）的环形建筑，它的外形是一个鼓。外圈的房间按照建筑的圆形轮廓顺势而建，但圆形建筑内部的大部分结构都是法院房间，其中最大的两个房间为常规的矩形。这座圆形建筑不分正面和背面，挑战了传统的建筑模式——其入口为一个矩形的透明玻璃幕墙，进入之后穿过一条通风的中央轴向通道，即可达到建筑尽头一个肖似入口的双高"大落地窗"。

最后，我们要介绍位于西班牙小镇科尔多瓦的科尔多瓦司法大楼（138页）（图13），该地区曾于中世纪被阿拉伯人统治过，因此建筑师借鉴了阿拉伯屏风的设计理念。同时，Mecanoo Architecten建筑师事务所和Ayesa工程公司在设计该建筑时，也针对其城市环境做出了相应的设计。在该建筑案例中，建筑师从联合国教科文组织东南部世界遗产中心发现的该城市的本土建筑中汲取了设计灵感，同时该建筑非常庞大，足以与邻近的普通高密度公寓楼形成对比。在1236年之前的五个世纪里，科尔多瓦一直处于伊斯兰的统治之下，曾是

by massive social housing. As such, it plays a role in the city's Grand Paris urban agenda to bridge the social and geographical gap across the Périphérique. The legal profession was reluctant to move out from the old law courts near Notre Dame to what was a soulless urban edge zone, but as Clichy regenerates, Paris should hear less grumbling from them.

Architect Phillipe Gazeau had no urban context to address at the site of his Palace of Justice, which lies below a wooded hill at the edge of the small southern French town of Foix (p.172) figure 12. The most notable element is the sinuous ramp that leads up to a floating circular building that touches the hill behind. The building form is not a ring like, for example, Foster + Partners's Apple Park, Cupertino, California (under construction) figure 11, but a drum. The perimeter rooms are shaped by the circumference, but most of the interior of the circle is built structure housing court rooms, the largest two of which are conventionally rectangular. Nor does the circle challenge the paradigm that buildings must have a front and back – a rectangularly framed curtain wall of clear glazing marks the entrance, leading into an airy central axial passage which ends in a similar double-height "picture window".

Our last courthouse takes us to Spanish town with a history of Arab rule in the Middle Ages, and the architects borrow from Arab ideas of permeable screens like the moucharabiya.

Palace of Justice (p.138) in Córdoba, Spain by Mecanoo Architecten + AYESA figure 13, is also designed to react to its urban context. It takes inspiration from the city's vernacular architecture found in the central UNESCO World Heritage

©Fernando Alda

13. 科尔多瓦司法大楼，Mecanoo Architecten建筑师事务所＋AYESA
13. Palace of Justice in Córdoba by Mecanoo + AYESA

©Mecanoo Architecten

14. 代尔夫特市政厅和火车站，Mecanoo Architecten建筑师事务所，2017年
14. Delft City Hall and Train Station by tMecanoo Architecten, 2017

世界上最大、学术造诣最高的城市之一。新建筑的玻璃纤维增强混凝土覆层上的孔洞有着伊斯兰屏风的影子。建筑外形为一个长块单体结构，其中一部分被垂直切割成带有悬臂的凹痕空隙，延伸到中央交通循环结构中。这些阴暗的空隙缓和了建筑厚重的质感，并对科尔多瓦的庭院进行了重新诠释。这些空隙内衬颜色较深的多孔陶瓷板，其作用与Mecanoo Architecten建筑师事务所设计的荷兰代尔夫特市政厅和火车站（2017年）（图14）的空隙相似，是它的一种变化形式。在图14中，大块建筑上的凹痕仿照别具一格的旧城镇景观的规模和纹理而建，但空隙仅存在于建筑上层。与利摩日法院相似，科尔多瓦法院的入口也是一个被遮蔽的切口缝隙，入口也是位于建筑一角，上方为悬臂结构。

我们这里所调查的当代建筑是否与最初的雅典法院有着共同之处？事实上，我们只能从废墟和其他类型的建筑中推断出雅典法院原来的样子。即便如此，我们仍可以很明显地看出，这些建筑无论从形式、结构还是材质方面都比古希腊建筑更加多样化。一些法院仍然沿用了前现代主义中的新古典主义风格，它们那纪念碑式的入口尤其能体现这一点。但是在建筑设计中，更多体现出的还是建筑的透明性，建筑师利用幕墙玻璃和复杂的墙体设计实现了这种透明性。与其说这些建筑借鉴了古希腊的建筑风格，倒不如说它们契合了古希腊精神——正义不是隐藏的，而是对所有人可见。

site to the southeast, and simultaneously the overall form is massive enough to counterpoint the undistinguished high-density apartment blocks adjacent to it. For five centuries until 1236, Cordoba was under Islamic rule, and temporarily became one of the world's largest and most scholarly cities. There is an echo of Islamic screens in the perforation of the glass-fibre reinforced concrete (GRC) cladding. The long monolithic block is partially cut vertically, creating indented voids which include cantilevers and reach into the central circulation spine. These shady voids temper the buildings' mass and re-interpret the courtyards of Cordoba. They are lined with darker perforated ceramic panels. The morphological role of the gaps is a variation on the Mecanoo Architecten's Delft City Hall and Train Station in the Netherlands (2017)[figure 14], where indents in a massive block reference the scale and grain of a very different old townscape, but there, the voids are only on the upper floors. The entrance in Cordoba is also via a shady incised void, just one storey high, which (as at Limoges) is made by cantilevering the building over a corner.

Do the contemporary buildings we survey resonate with the original Athenian courthouses? The fact is that we can only infer what they looked like from their ruins and the surviving architecture of other typologies. Nevertheless, it is clear that form, structure and materiality has diversified widely from ancient Greece. Some courthouses still have an echo of pre-modernist neoclassicalism, notably when they make monumental entrances. But transparency, enabled by curtain wall glazing and sophisticated screens, is more dominant. And that relates to the ideas rather than the architecture of Ancient Greece – that justice is not hidden, but for all.

拉里奥加法院
La Rioja Courthouse

Pesquera Ulargui Arquitectos

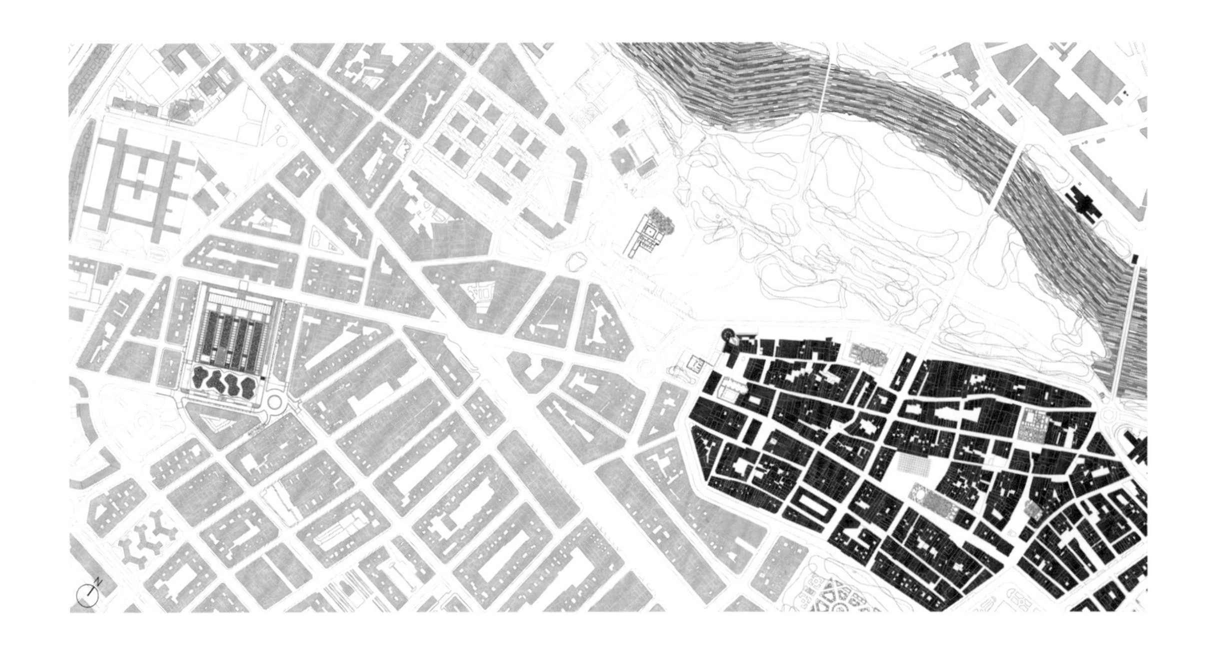

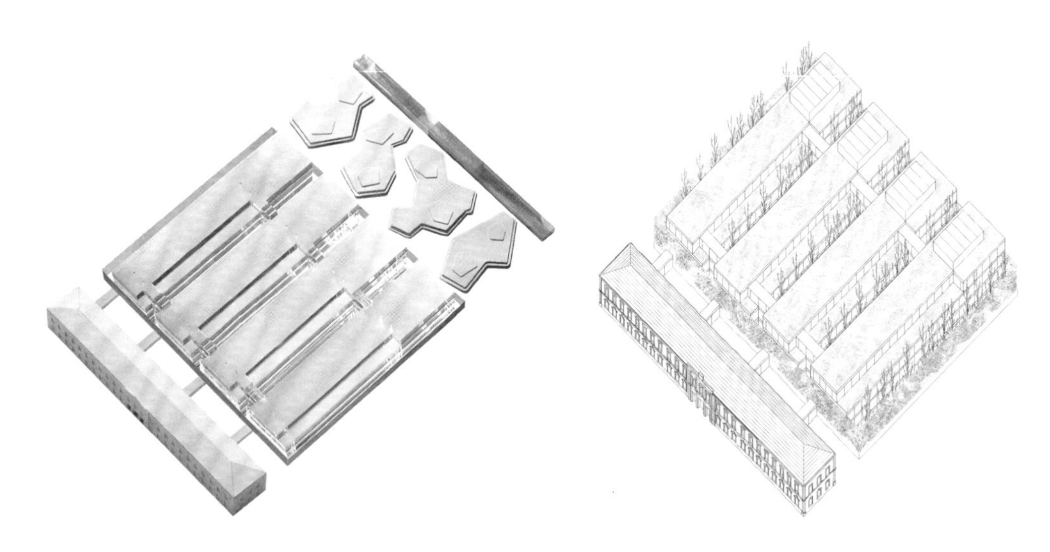

该建筑项目有两种相互关联的功能。首先我们应当将它理解为一座实用、经济实惠而又精确的建筑，符合一个精准的建筑项目所要求的限制条件。同时，它是一座宫殿，是一座代表国家对公民权力的建筑，其室内设计必须传达平等和责任，而且在某种程度上也传达了真理。建筑的这两种功能如何能相互协调？而效用与表现、经济与意义、效率与身份认同又如何能相互结合？

建筑师利用了旧建筑来达成这一目的。新的建筑会给人带来不安全感，而建筑师对旧建筑进行了重新利用，通过这些经年累月保存下来的建筑部分来安抚人们的内心。原来的楼阁被清空，变成了公共活动区域。大门确实是整个宫殿最重要的部分，因此在设计新建筑时，建筑师要严格地从经济性和功能性对其进行考虑。

建筑师利用干预的规模来治愈我们加速城市发展所带来的"未治愈的创伤"。建筑师力求在密集的建筑中开创出尽可能大的公共空间，在建筑中打造了一个新的城市大厅，将新鲜的氧气引入密集的居住环境中。建筑师用一个大型的植物外衣将四座新建筑结合在一起。植被外衣这个轻质元素调和了建筑方案中功能性和此类行动产生的城市冲突之间的矛盾。"树叶包裹内容"是对建筑师意图的最直接的诠释。同时，这种以植物为面纱的建筑体量使人们无法从街上直接看到法庭内部，为大楼创造了一个阴凉清新的空间。

The project has two related functions. It must be understood as a useful, economic and accurate construction complying with the restrictive conditions of a very precise program. At the same time, it is a palace, a building that represents the power of the state in relation to its citizens. The interior must convey equality, responsibility and, to some extent, the truth. How can the building's two functions be reconciled? How can utility be combined with representation, economy with meaning, and efficiency with identification?

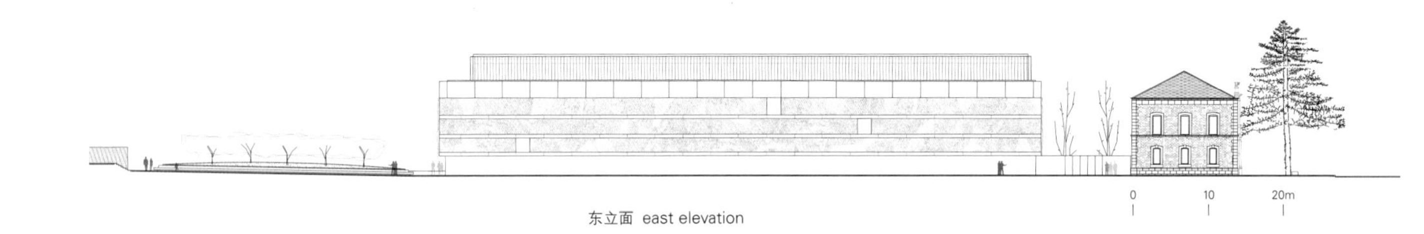

东立面 east elevation

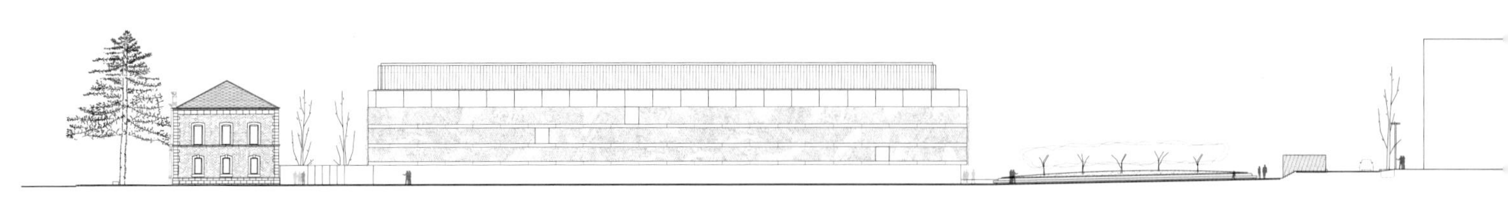

西立面 west elevation

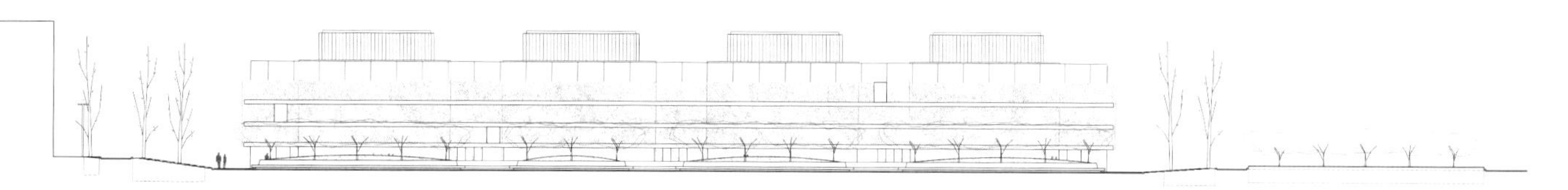

南立面 south elevation

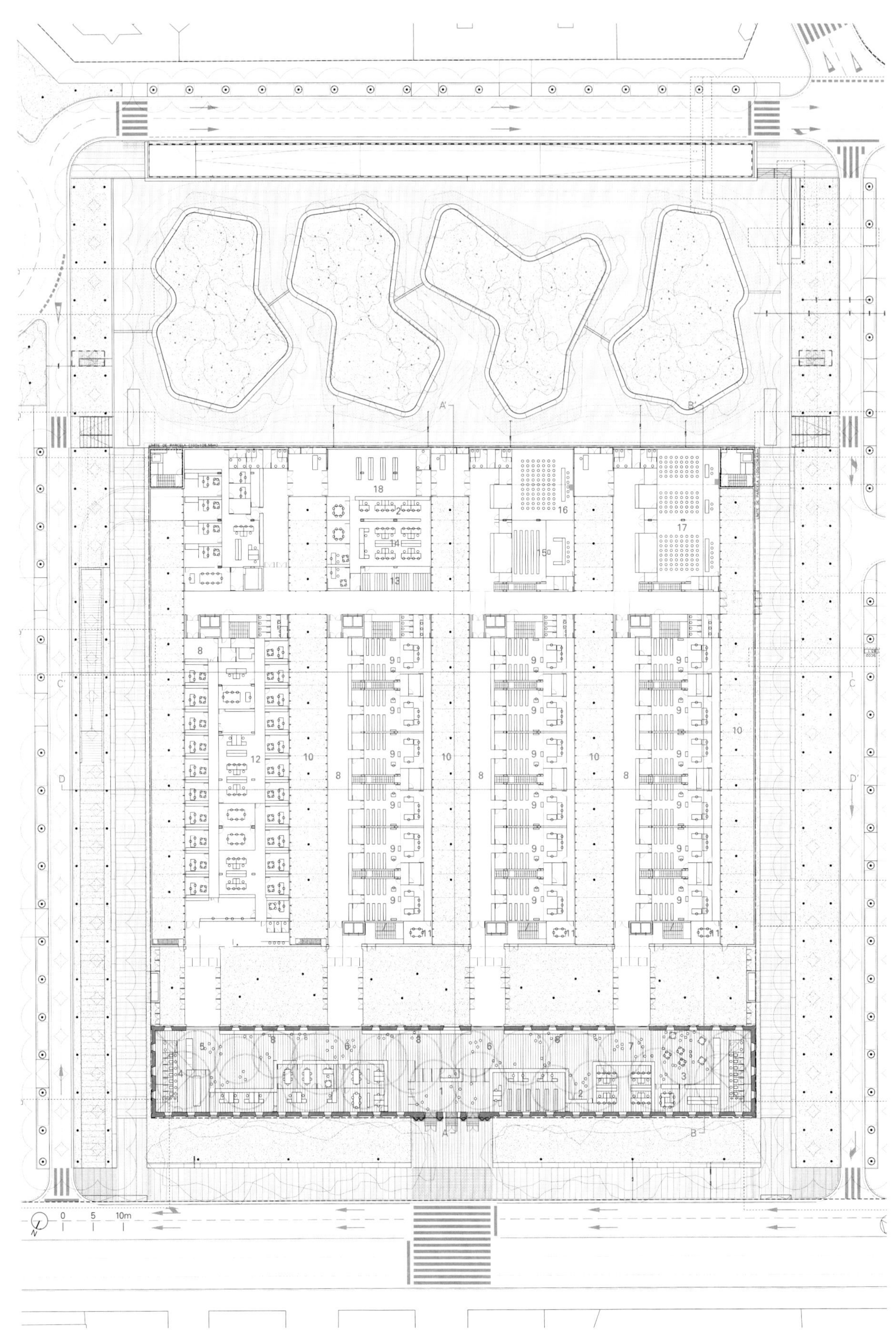

1. 入口 2. 信息处/登记处 3. 咖啡厅 4. 卫生间 5. 入口社交服务设施 6. 调查庭用的房间入口 7. 未来建筑入口 8. 等待区 9. 调查庭用的房间 10. 天井
11. 会议室 12. 地下庭院 13. 档案室 14. 办公室 15. 省级听证厅会议室 16. 婚姻注册室 17. 多功能室 18. 户籍办公室
1. entrance 2. info / registration desk 3. cafe 4. toilets 5. access social service 6. hearing room entrance 7. future building entrance 8. waiting area 9. hearing room 10. patio
11. meeting room 12. underaged court 13. archive 14. office 15. provincial hearing room assembly hall 16. wedding room 17. multipurpose room 18. civil registration
一层 ground floor

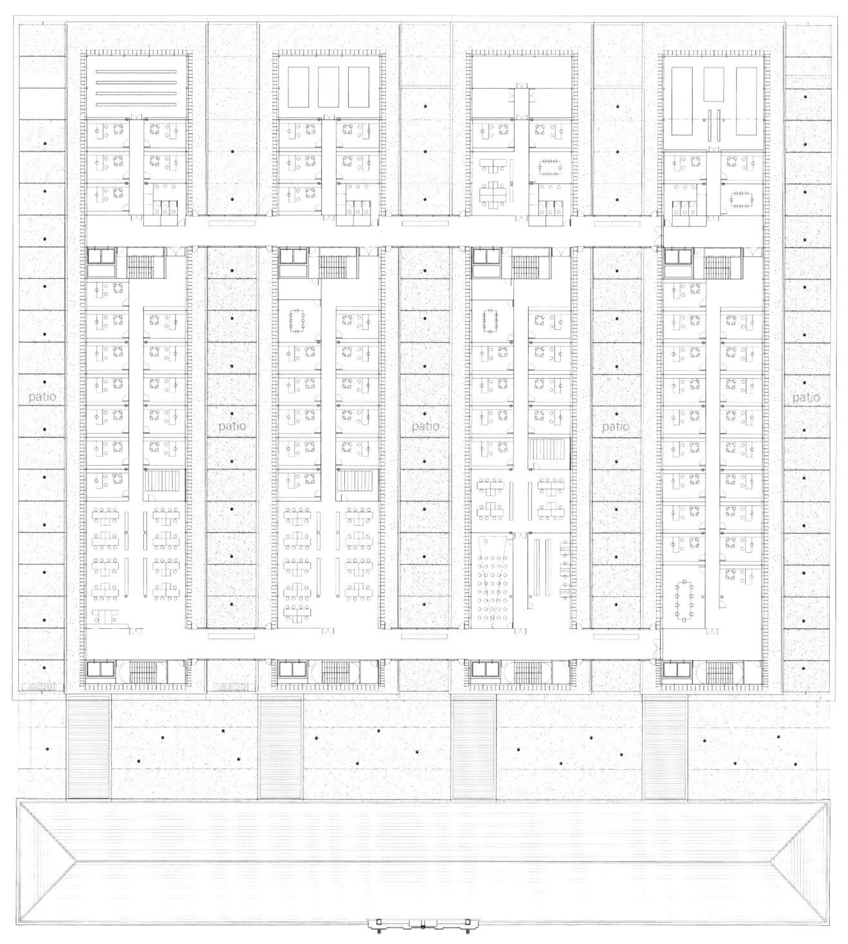

四层 third floor

1. 信息处 2. 天井 3. 休闲花园 1. info desk 2. patio 3. garden to relax
二层 first floor

To do this, the architects used the old building. Its reuse brings to the fore the psychological stability of what has remained over time in the face of the insecurity generated by the construction of a new building. This pavilion is emptied to transform it into a public place. The great door of the palace is really the most significant space of the whole. And this allows the new constructions to be conceived from strict conditions of economy and functionality.
The size of the intervention has been exploited to heal the "unhealed wounds" that our accelerated urban development always entails. The work on the tight building has been reflected in the search for the largest possible public space, creating a new urban hall that breathes fresh oxygen into the dense residential fabric of the environment. A large plant wrap unifies the four new buildings into one. We pretend that this light element mediates between the functionality of the proposal and the urban conflicts that such action generates. "Wrapping content in leaves" would be the most direct explanation of the architects' intention. At the same time, this vegetal veil blocks direct views of the courtrooms from the street, creating a shaded and refreshing space for the building.

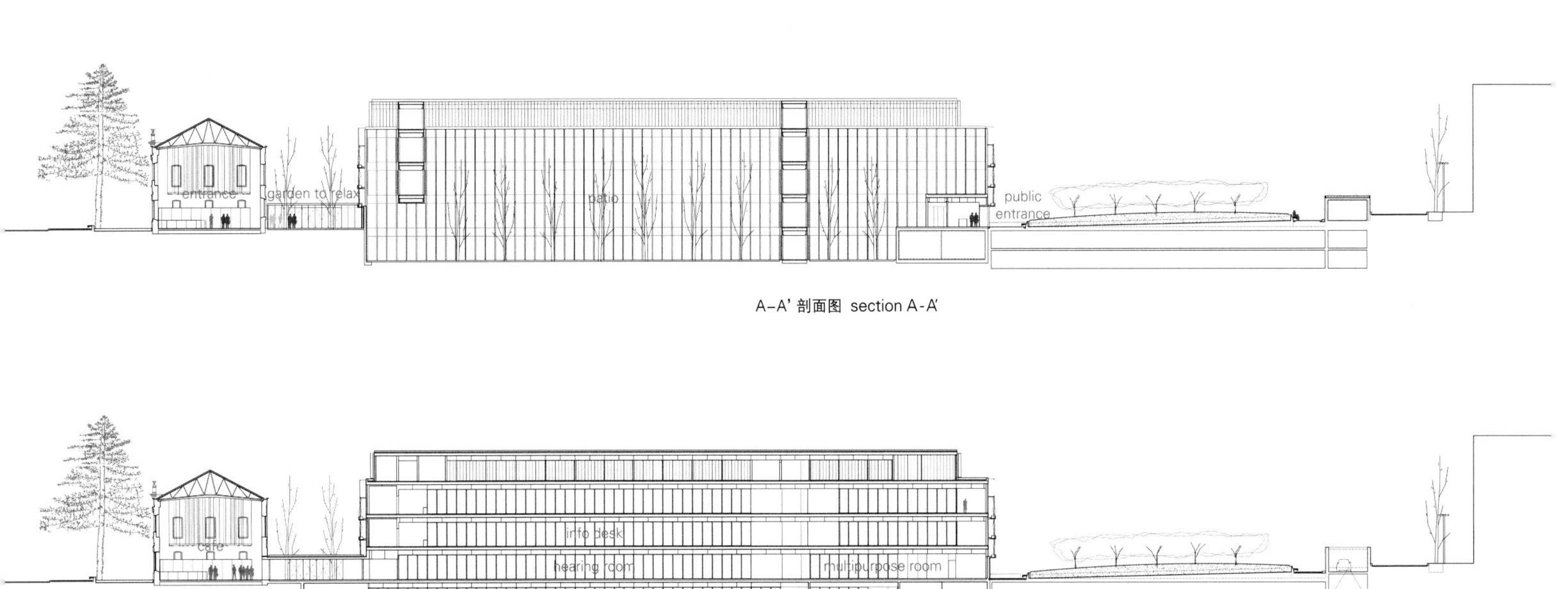

A–A' 剖面图 section A-A'

B–B' 剖面图 section B-B'

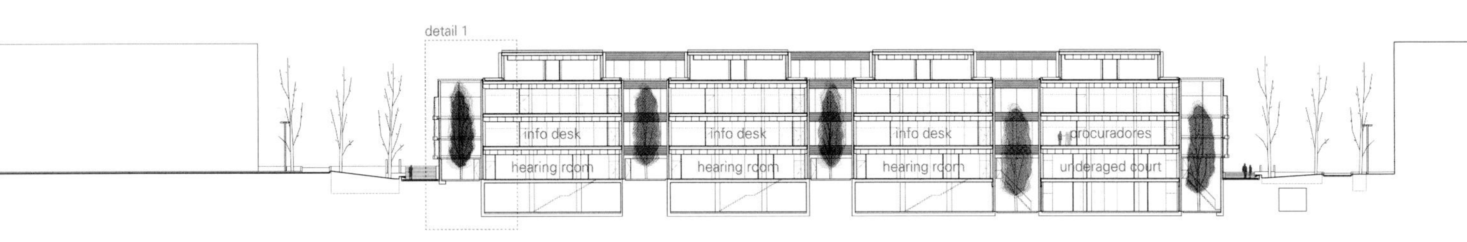

C–C' 剖面图 section C-C'

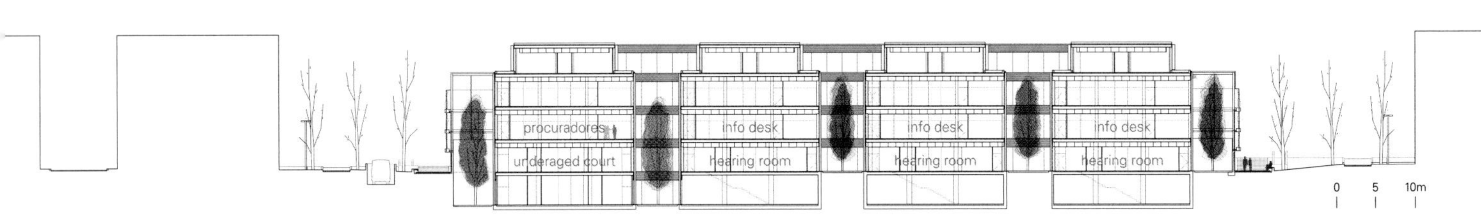

D–D' 剖面图 section D-D'

项目名称：La Rioja Courthouse
地点：C/ Padre Amoedo esq C/ Sierra, Logroño, La Rioja, Spain
建筑师：Pesquera Ulargui arquitectos slp-Jesús Ulargui Agurruza, Eduardo Pesquera González
合作方：Jorge Sánchez Limón, architect; Javier Mosquera González, architect;
Judith Sastre Arce, architect; Jose Antonio Vilches Menéndez, civil engineer;
Euteca sl, structural engineers; JG & Asociados, CVSE engineers; Miguel Fernández Rueda, architect;
Araceli Barrio Fernández, architect; Mª de las Nieves Vicente Recio, architect
承包商：Obra Civil Euroservicios, Aransa Construcciones, Sacyr Construcción
客户：Gobierno de La Rioja
建筑面积：28,989m² / 竞赛时间：2011 / 施工时间：2012—2016
摄影师：©Pedro Pegenaute (courtesy of the architect)

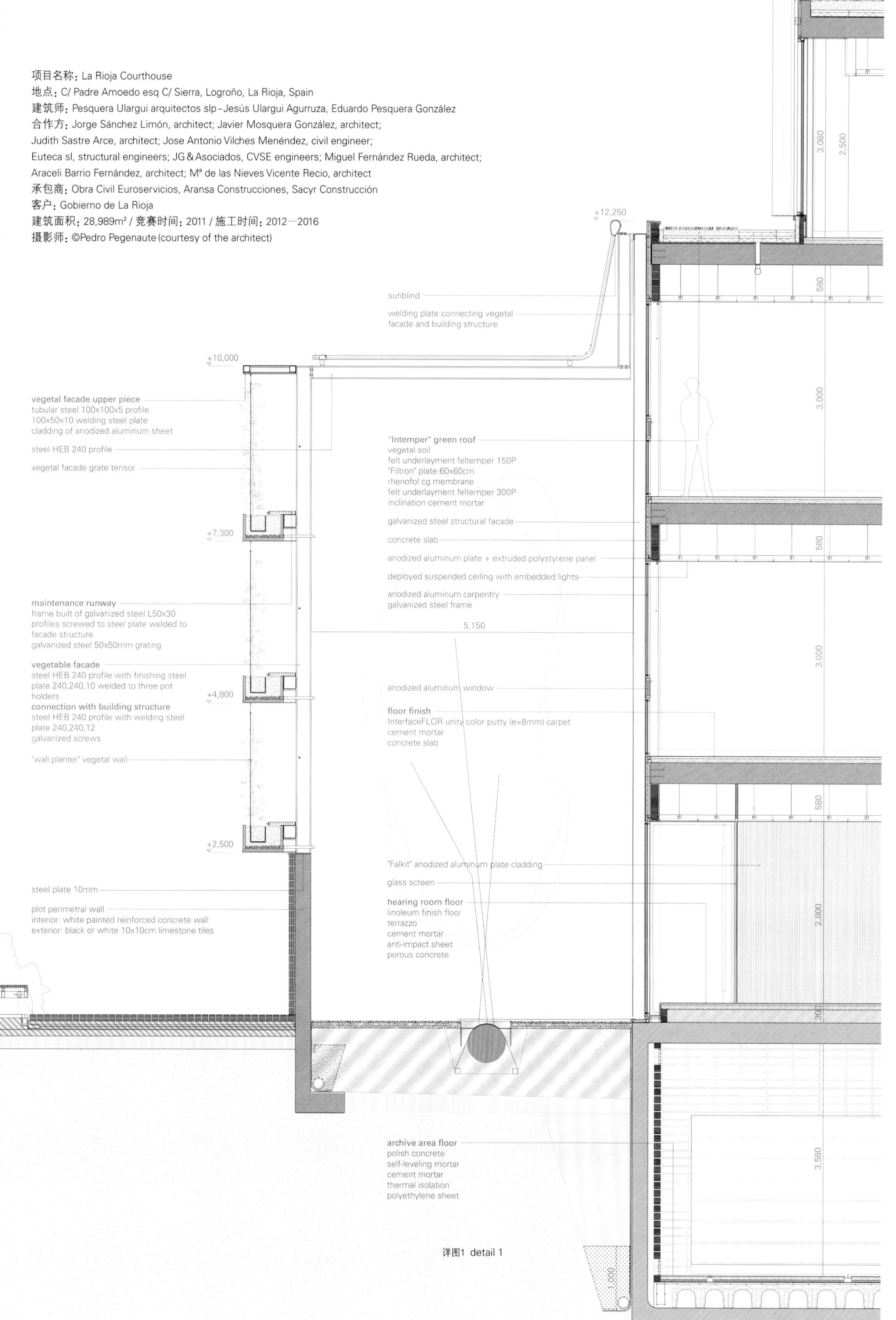

详图1 detail 1

调节所产生的稳定感和安全感

Sense of Stability and Security by Moderators

Interview with Pesquera Ulargui Arquitectos

C3 拉里奥加法院的旧建筑最初是为何而建？您决定保持建筑原状的主要原因是什么？

Pesquera Ulargui Arquitectos（下面简称"PUA"）主要的设计决策就是保留前军事医院的第一座楼阁作为"机构的正面"。该楼阁的主要价值在于其历史悠久，能够唤起市民的记忆。毫无疑问，重新利用该楼阁将有助于保持建筑整体的"稳定性"，消除新建筑给人们带来的"不安全感"。

该方案通过移除现有的建筑结构，将楼阁的内部完全清空，并将它改造成优质的公共空间，作为"宫殿"大门，成为整个建筑的重要部分。门厅既可以用于发布信息，也是所有公共访客的等候区。从门厅出发，有专用的路线通往各个功能区，最大限度地减少了建筑其他区域的人员聚集。

C3 旧建筑保留了原有的结构和几何形状，被改造成了法院的入口，成了城市和法院之间的缓冲区。通过这种设计策略，您希望达到什么样的目的？在这个过程中，您有没有遇到意想不到的挑战？

PUA 这个项目需要满足两个条件。一方面，首先我们应当将它理解为一座实用、经济实惠而又精确的建筑，符合一个精准的建筑项目所要求的限制条件；另一方面，它必须符合一个"宫殿"的定位，这类建筑代表国家公共权力，这是其对公民的承诺。它必须传达平等、责任，以及在某种意义上真理统治一切的理念。

C3 法院的新建部分使用了玻璃和铝材，营造出一种明亮、透明和轻盈的感觉，这与保留的旧建筑给人的坚实和墓碑般的感觉截然相反。您在设计中运用如此强烈的对比，目的是什么？

PUA 如果保存下来的楼阁是用来塑造建筑特点的，那么新建的部分则是从经济条件和功能方面的严格考量。新建筑在组织、测量和建造方面都按照其应有的功能进行精确调整。与一个大机构比起来，

C3 *What was the old building of La Rioja Courthouse originally built for and what was the main reason for your decision to maintain it as it was?*

Pesquera Ulargui Arquitectos The principal decision was to maintain the first pavilion of the former Military Hospital as the "institutional front". The main value of the existing pavilion lies in its permanence in time, which makes it appear in the memory of citizens. There is no doubt that its reuse will contribute to the overall "stability" of what remains, opposing the "insecurity" that the new buildings bring.

The proposal was to completely empty the building by removing the existing structure and to transform it into the public place par excellence in the great door of the "Palace", the significant space of the whole. The lobby aims to become the place for information distribution, the waiting area for all public visits, and space from which the most specialized routes depart to minimizes the accumulation of people in the rest of the building.

C3 *Maintaining its original structure and geometry, the old building was transformed into the entrance to the courthouse, acting as a buffer area between the city and the courthouse. What did you expect to achieve from this strategy? Did you encounter any unexpected challenges along the way?*

PUA The proposed building faces double conditions – on the one hand, it should be understood as a useful, economic and exact construction to respond to restrictive conditions of a very precise program; on the other, it has to respond to its status of "Palace", with the commitment of this type of buildings has with the citizens as a representation of the public power of the State. It has to convey equity, responsibility and, to a certain extent, truth must rule.

C3 *The new part of the courthouse is finished with glass and aluminum, creating a bright, transparent and light feel, contrary to the solid and grave-like feel that the old part gives. What was your intention for such a strong contrast?*

PUA If the existing pavilion is responsible for shaping the character, the new constructions are conceived from strict conditions of economy and functionality. They are "what they should be" in a precise adjustment of the organization, measurement, and construction. A system of the new judicial programs correspond more to an administrative office than to an institution - it is about serving the needs rather

一个行政办公室与新的司法项目系统更加匹配——它的功能是服务于民众的需求，而不是制造麻烦。因此，作为一个行政办公系统，新建筑采用了这类建筑常用的铝材和玻璃表面，以确保为其用户提供最佳的工作条件。

C3 您为什么要把法院的新建部分建成四栋楼，并将它们置于旧建筑的垂直方向？

PUA 我们设计了一系列相互连接的建筑，经过调整，其间距允许自然光线进入。我们按照距离和用途将这些建筑的功能组织（可随时修改）按照顺序排列：从最为公开的听证室到私人的司法办公室。该建筑系统通过两条横向路线的连接，以便根据未来的需要在随后的独立阶段对建筑综合体进行施工建造。

C3 新法院的外墙被植物包裹，庭院置于建筑之间。在建筑中引入这些自然元素的目的是什么？请您多和我讲讲设计中的整体城市概念，包括您设计的法院南端的公共广场。

PUA 尽管我们设计了具有活力的建筑设施，但我们仍提议建造一个大的植被墙衣，将四座新建筑合为一体，其高度与旧的军事展馆高度相当。该墙衣由张拉钢筋网制成,其表面将种植不同种类的攀援植物，会随着季节的变化而改变颜色。这种植物面纱使人们无法从街上直接看到法庭内部，为大楼创造出一个阴凉清新的空间，并向这个城市展现出正义、人性和友善的一面。

谈到城市概念，我们在楼层的数量和楼间距方面尽量压缩空间，以留出更多的公共空间。我们设计了这个新的城市大厅，也是为了日后建立地下停车场以及一个小公园。这个小公园将由花床构成，随着季节的变换更换不同的芳香花卉植物，从周围的所有建筑上都可以看到这幅巨大的花卉图画。花园中设有大长椅，供人们等候、冥想、休息使用，而花园中的空隙为周围居民提供游戏和居民服务的场所。

than creating new problems. Therefore, as an administrative office system, the new buildings adopt the aluminum and glass finishings typical of this type of buildings in order to ensure the best working conditions.

C3 *What made you divide the new part of the courthouse into four buildings and laid them out perpendicular to the old part of the courthouse?*

PUA We proposed a serialization of interconnected buildings whose adjusted equidistance allows the entry of natural light and whose program organization (which could be modified at any time) is in the sequence of distances and uses: from the most public (hearing rooms) to the private (judicial offices). Linked by two transversal routes, the system easily manages to arrange the construction of the complex in successive independent phases according to future needs.

C3 *The outer walls of the new courthouse are wrapped with plants and the patios are placed between the buildings. What was the purpose of introducing such natural elements in the building? Please tell us more about the overall urban concept including the public square at the south end of the courthouse.*

PUA Regardless of the active design of the facilities, we suggested creating a large vegetable wrapper that unifies the four new buildings into a single volume and whose height relates to the scale of the old military pavilion. Composed of a tensioned steel mesh, on its surface will grow different climbing species that will alternate their coloration throughout the seasons. This vegetation veil will prevent direct views of the courts from the street, promote shadow spaces and refreshments for the building, and project towards the city of a human and a kind face of justice.

Talking about the urban concept, the strategy of compacting the program as much as possible, both in the number of floors and in the distance between the buildings, resulted in a larger public space. We planned this new urban hall for a possible future construction of an underground parking lot with a small park on it. Made up of a single flower bed, the interiors of aromatic plants will alternate with seasonal flowers, creating a large drawing that can be seen from all the surrounding buildings. The garden offers large benches for waiting, contemplation and rest, whose interstices provide spaces for games and services to the neighborhood.

里摩日法院
Limoges Courthouse

ANMA

地区法院（TGI）和里摩日诉讼与贸易中心现在被整合到了一座新建的法院大楼中，并于2016年6月向公众开放。地区法院大楼好几处位置已经破败，难以实现团结和凝聚力。现在，建筑师将它们联合安置到里摩日法院大楼里，以恢复其一致性和往日的工作效率。里摩日法院庄严地矗立在温斯顿丘吉尔广场的一角，靠近历史悠久的艾因宫广场，在这座广场上还保留有上诉法庭，地区行政管理局也于6月搬入该建筑（在TGI迁至新大楼之后）。因此，这座舒适的现代化新大楼为法院全体官员和治安官带来了舒适和便利，让他们能够更好地完成使命，实现诉讼当事人的期望。

这座白色的立方体建筑面向城市开放，在肯定民主价值观和接受度的同时，体现了君主的地位和法治国家的力量。建筑的前院是一个单纯的体量，巨大的白色饰板遮挡其上，就像一个巨大的树冠在俯瞰其下的公共空间。两个垂直面由大而薄的水平板连接，沿对角线方向在广场上行走时显得视野开阔，并围合了一方天空，以及被植被覆盖的露台。因此，巨大而坚固的正义宫，在广场的空旷、前院的体量和朝向天空打开的窗户之间建立了强有力的联系。

高贵感十足的建筑立面由带有渐变扭曲的竖框构成，象征着司法制度。这种扭曲是为了保护建筑的玻璃表面，防止阳光直射（尤其是在高楼层的办公室），以及按照建筑师的设计将人们的视线引向一个特定的视角。特别是监狱所在的东南立面，扭曲的表面将人们的视线引向温斯顿·丘吉尔广场，避免了直接看到监狱的墙壁。立面上的这些垂直构件采用了多孔而又坚固的材质，反衬出了白色光滑的庭院体量。

前院直接通向内部房间，人们不会迷失方向。走廊就像一个双层高的画廊，通往礼堂的入口显得尤为突出。玻璃天花板和一个光圈结构使人们能够直接透视到花园底部，并将光线引入室内。因此，阴影和光线交替出现在长长的画廊上，产生了一种万花筒般的效果，使这条通道充满戏剧色彩。由棱镜构成的光圈结构将外部景观神奇地围合在其中。

法院建筑的总体规划是清晰的，功能性和演变性都很强。它满足了功能要求，优化了视线和场地配置。它考虑了刑事和民事领域的划分，尊重公共空间和私人空间的分离。审判室分布在一层台阶走廊两侧。地方法官的出入口是固定在外围的，而办公室则设在画廊S形建筑物的楼上。建筑侧面点缀着大量植物，形成了可以进入的露台和花园，让人们得到视觉上的休息。

The district courts (TGI), instance and trade Limoges are now grouped in a new courthouse and opened to the public in June 2016. The courts had been broken up on several sites, which made difficult of attaining cohesion and solidarity. Joining them together into Limoges Courthouse aimed to gain consistency and efficiency back. Solemnly marking the corner of Winston Churchill Square, it is close to the historic Palace of Aine Square, which retains the Court of Appeal, joined in June by the Regional Administrative Service (following the move from the TGI to the new building). Officials and magistrates of all the courts thus benefit from a modern and comfortable building, at the height of their missions and the expectations of the litigants.

The cubic volume that is white and open to the city, expresses a regal place and the power of the state of law while affirming the democratic values and reception. The forecourt forms a pure volume, sheltered by the large white

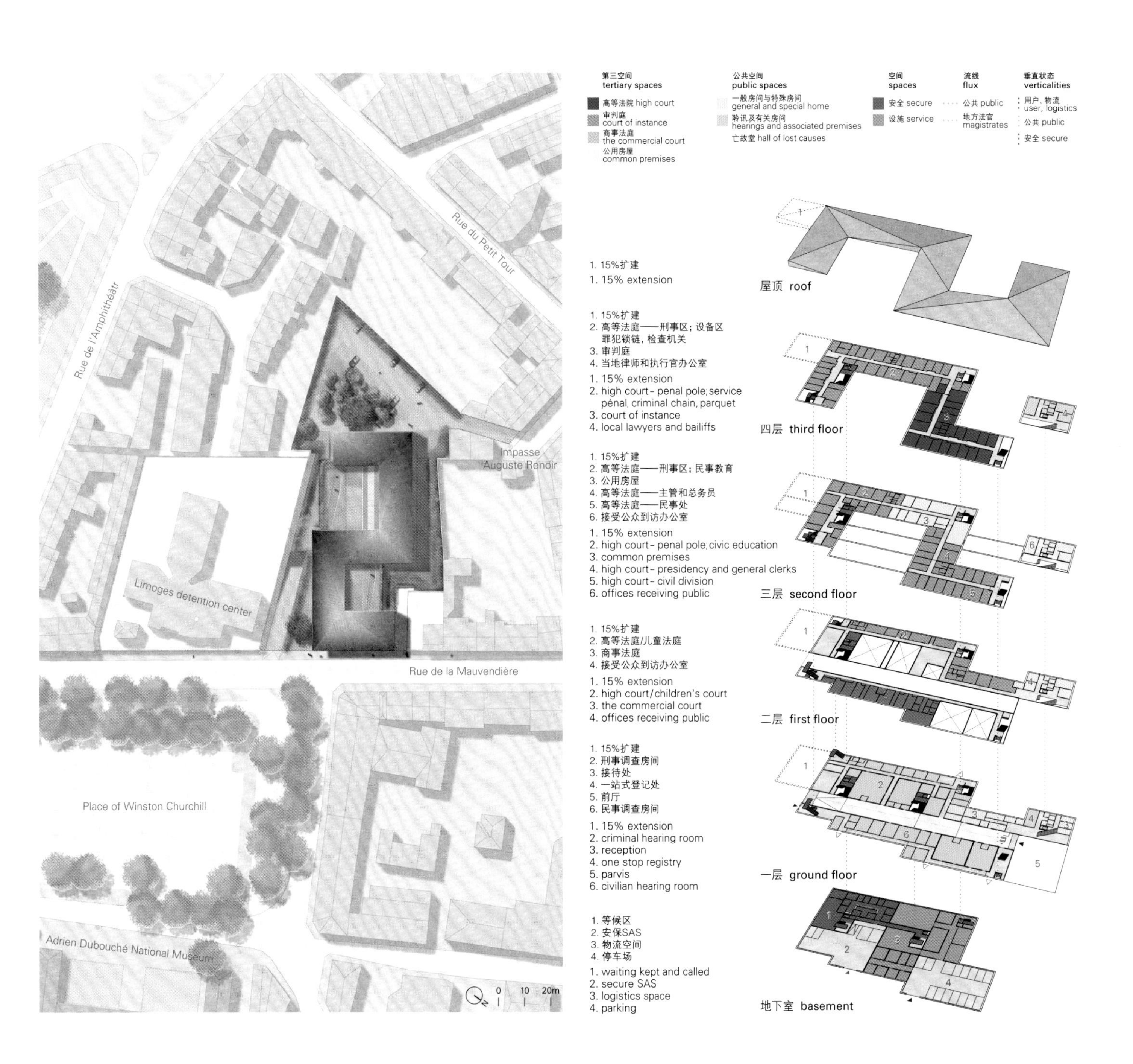

plaque which resembles a monumental canopy that overlooks the public space. The two vertical surfaces, connected by the large and thin horizontal plate, free the vision while walking diagonally from the square and frame a fragment of the sky and the vegetated terrace. Thus, The Palace of Justice, monumental and firm, creates a strong connection between the emptiness of the square, the volume of the forecourt, and the window to the sky.

The noble facades, emblematic of the institution of Justice, are created by vertical mullions set with a progressive twist. The twisting stemmed from the will to partially protect the glazed surfaces from direct sunlight (especially at the upper floor of the offices) and the will to direct the eye towards a privileged view. In particular, on the south-east facade, where the prison is located inside, the surface directs the sight towards the Winston Churchill Place, avoiding the direct view on the carceral wall. The porous and solid materiality of these vertical elements of the facade emphasizes, by contrast, the white and smooth volume of the court.

The forecourt directly leads to the room of not lost. The corridor serves like a long gallery in double height while punctuating the access to the auditoriums. Glass ceiling and an optical ring allow a perspective view to the bottom of the garden and stages the light inside. An alternation of shadow and light thus lands over the length of the gallery, producing a kaleidoscopic effect that theatricalizes the path. This optic ring made of prisms frames the exterior landscape in a magical end of course.

The general plan of the palace is clear, functional and evolutive. It meets the constraints of the program and optimizes the views and the configuration of the plot. It considered the partitions of the domains of the criminal and the civil, to respect the separation of the spaces accessible by the public or that reserved to the private. The courtrooms are distributed by the gallery of steps on the ground floor. The magistrates' access is secured on the periphery while the offices are placed on the upper floor, in a body of S-shaped building on the gallery. The plant masses punctuated on the side became the accessible terraces and gardens that provide a visual break.

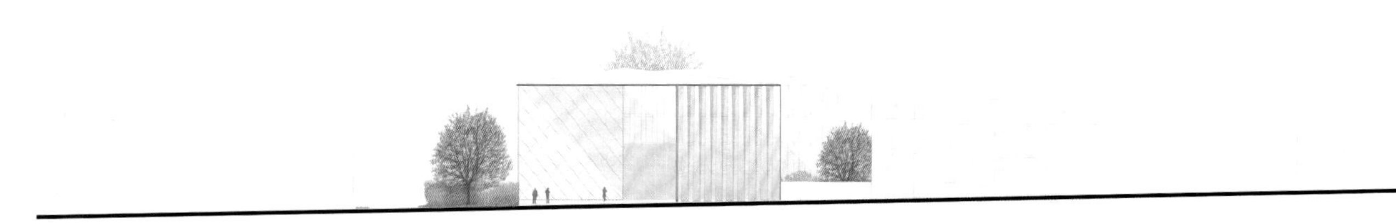

东北立面 north-east elevation

西南立面 south-west elevation

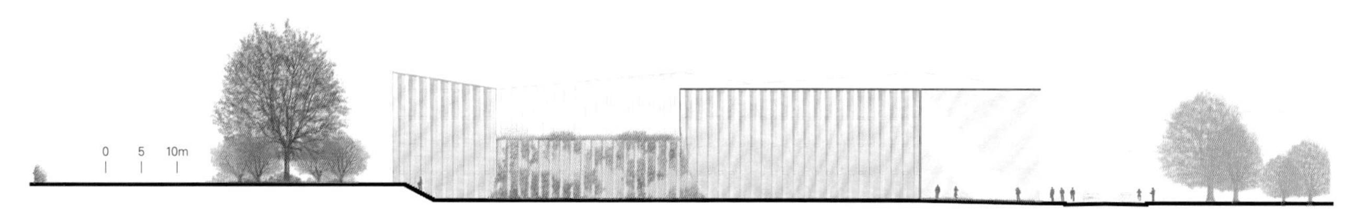

东南立面 south-east elevation

CITÉ JUDICIAIRE

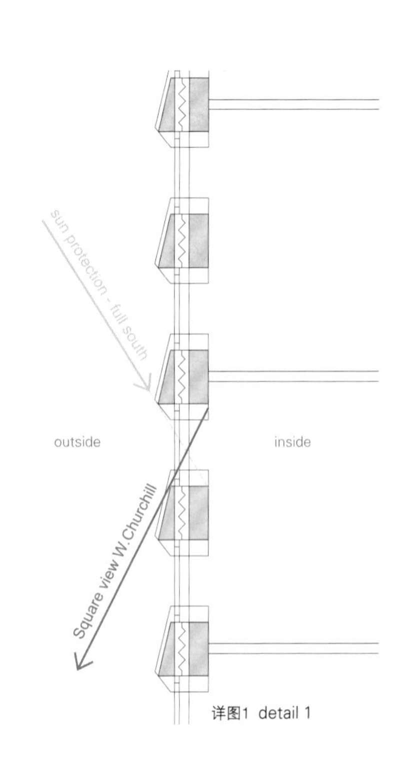
sun protection - full south
outside
inside
Square view W.Churchill
详图1 detail 1

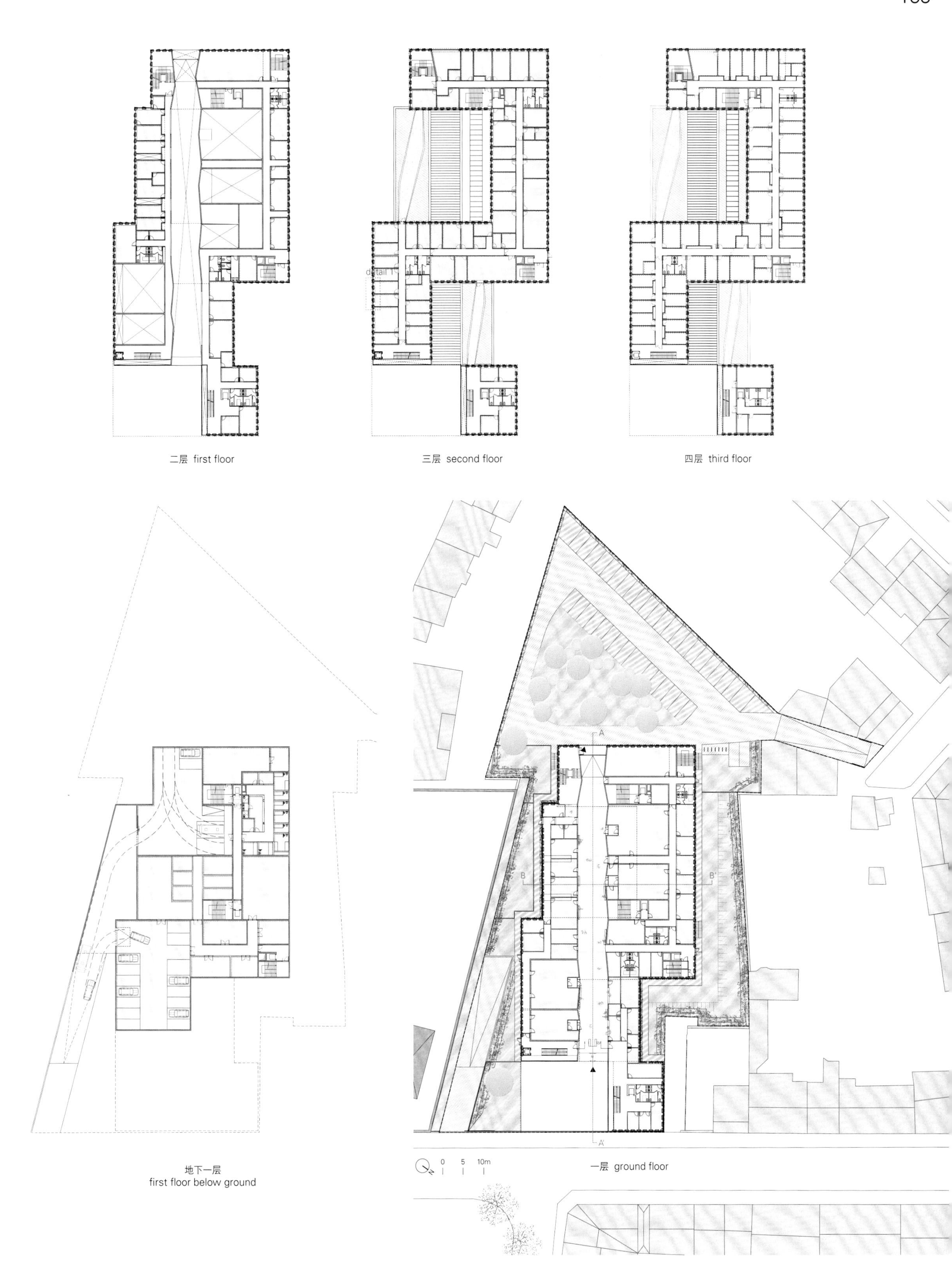

二层 first floor

三层 second floor

四层 third floor

地下一层
first floor below ground

一层 ground floor

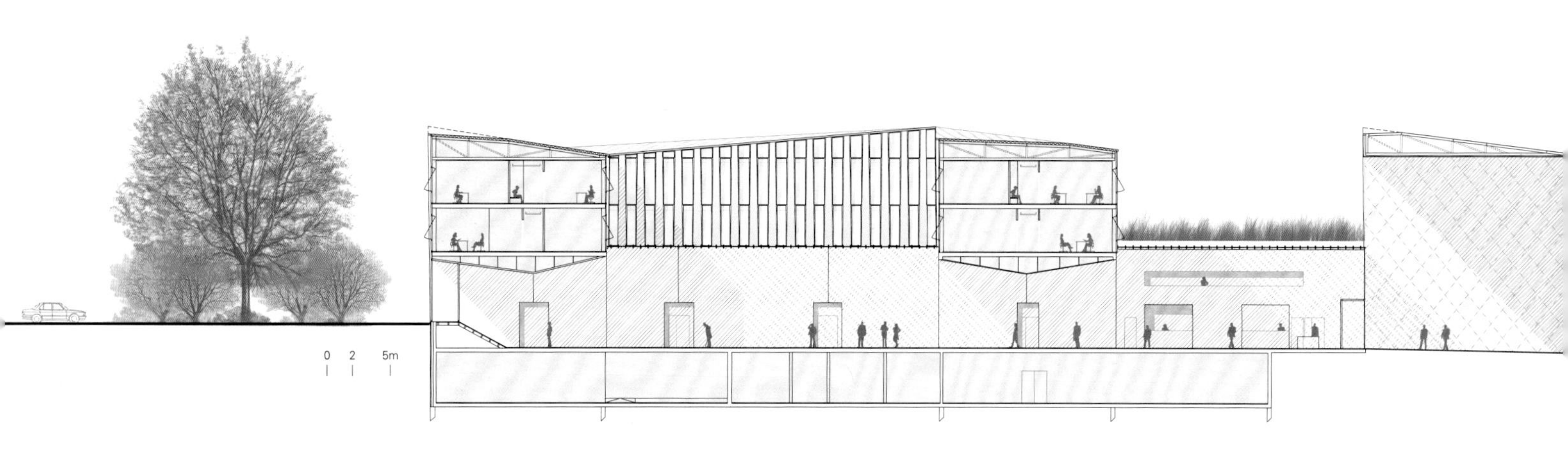

A-A' 剖面图 section A-A'

项目名称：Courthouse / 地点：23 Place Winston Churchill, 87000 Limoges, France / 建筑师：ANMA-Agence Nicolas Michelin & Associés
主持建筑师：Nicolas Michelin, Cyril Trétout, Michel Delplace / 项目团队：Hélène Galifer, Alice Perugini, Emmanuel Vinet, Guillaume Ribay, Juliette Chourrout, Iliana Genova, Melaine David, Emilien Lavice, Adèle Clin-Cassagne, Kevin Viel, Olivia Samit, Vivien Corre, David Cote / BET质量标准认证：DEERNS France

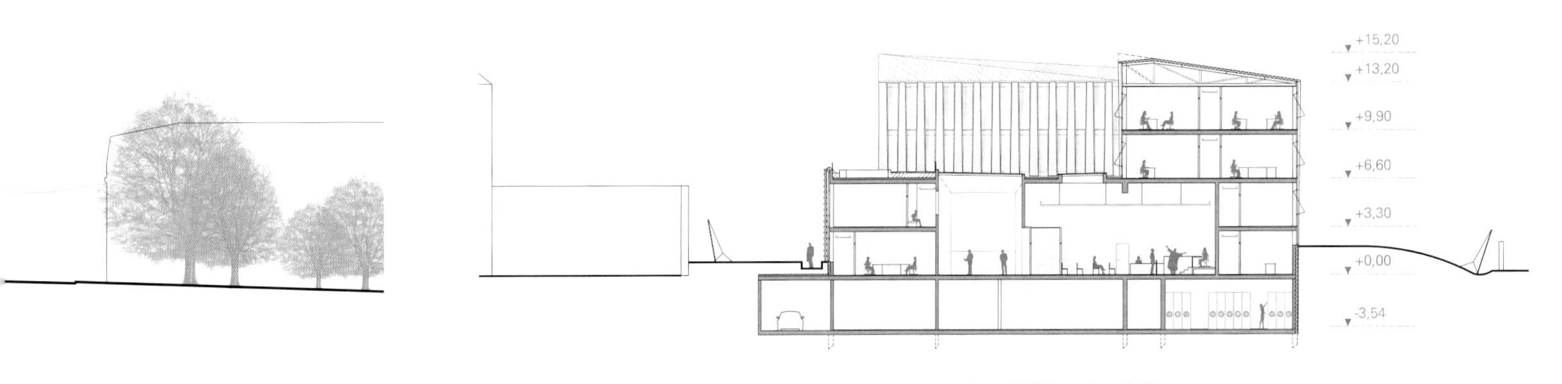

B-B' 剖面图 section B-B'

BET结构：BATISERF Ingénierie / 声学：PEUTZ&Associés / 维修：QUADRIM / 人工照明：8'18 / 外围护结构：SECMA / 经济学家：Bureau Michel FORGUE / 透视图：The Nood (Maison Générale) / 标志设计：Agathe DESOMBRES / 客户：APIJ-Agence Publique pour l'Immobilier de la Justice / 用地面积：6,144m² / 建筑面积：5,919m² / 总建筑面积：7,643m² / 竣工时间：2016 / 摄影师：©Sergio Grazia (courtesy of the architect)

科尔多瓦司法大楼
Palace of Justice in Córdoba

Mecanoo Architecten + AYESA

科尔多瓦的新司法大楼位于该区西部阿罗约·德尔·莫罗区一处平淡无奇的住宅区，这里是21世纪西班牙迅速城市化的产物。作为该区域城市结构特征的代表，这些住宅楼无法为城市带来公共空间或任何新奇的事物，但它们共同构成了紧凑而连贯的城市特色。在这个地区新建公共机构会使公共领域得到完善和升级，并为这个较新的社区赋予了更多公民性质。而司法大楼的设计概念就是要建筑肩负起这种强化公共认同的责任。科尔多瓦市中心位于该地的东南方，建筑体量较为紧缩，形成了一个宽敞的入口广场，与现有的沃尔塔·德尔·索迪洛花园相连。

临街的立面均采用几何图案的白色多孔混凝土板覆盖，在房间内产生出不同的光线效果。相比之下，露台的墙壁显得格外醒目，青铜陶瓷板墙壁就如同一张精致的面纱。

该项目的设计策略是通过片段式的体量形成一体化的城市形态。在设计项目时，建筑师遵循了与中世纪城市自主发展历程相似的策略：建筑的体量经过了精心雕刻，从而能够与周围环境相适应。最终建筑结构呈现出一种类似于拼图的造型，暗示了其构成的过程，同时模仿了科尔多瓦历史城区的密集肌理。

垂直裂缝在建筑中形成了一系列中庭，与当地常见的庭院类型相呼应。这些裂缝为这座大型建筑的中心区提供了自然光线和通风。可以说，这座建筑的可持续性不是靠昂贵的技术机械设备来实现的，而是通过对本土建筑的智能化诠释来实现的。

该建筑从街道层抬升2m，可通过倾斜的入口广场进入。抬升的体量赋予了建筑象征意义上的权力，同时解决了半开放式中庭带来的私密性和安全性的问题。由于该建筑被划分为几个独立的部门，因此需要基于外部和内部的空间层级秩序来修建多个入口。人们可以直接从主广场进入大楼，也可以从精心布置于其他三个立面的小型开放式中庭进入。

从主入口开始，室内布局显得十分清晰。位于中央的主厅提供了连接不同功能空间的交通循环轴线，通过横跨几个楼层来连接每个部门。同时它还将公共交通空间与外部庭院联系起来。主厅与外部的建筑遥相呼应，利用多层次的日光照射的空隙在整座建筑营造了如雕塑一般的中庭。

随着楼层的升高，内部功能空间的私密性也随之得到了提升。建筑的广场层为开放性地面层，设有审讯室、婚姻登记处和餐厅等公共区域。高级安全办公室位于上层庭院的周围，档案馆和监狱牢房则位于地下楼层。

The new Palace of Justice is located in Arroyo del Moro, a western district of Córdoba, which is dominated by anonymous housing blocks, the products of the rapid urban development of 21st-century Spanish cities. The blocks that characterize the urban fabric of the zone were not capable of generating public space or offering something new to the city, but collectively they form a compact and coherent urban identity. The addition of a public institution to the area creates the opportunity to upgrade the public realm and adds a civic quality to this relatively new neighborhood. This responsibility to strengthen the public identity informed the concept of the Palace of Justice. While the Córdoba city center is located south-east from the site, the building volume was condensed to create a generous entrance square to the north which connects to the existing Huerta del Sordillo gardens.

The streetside facades are all clad in white perforated concrete panels with a geometric pattern, thus producing varied light effects in the rooms. The walls of the patios stand out, by contrast, featuring bronze ceramic panels like a delicate veil.

The massing strategy creates urban integration through fragmentation. It follows a similar strategy to the spontaneous growth process of medieval cities resulting in a volume which is carefully sculpted to adapt to the surrounding context. This results in a puzzle-like structure which hints at the process of formation and emulates the experience of the dense historical center of Córdoba.

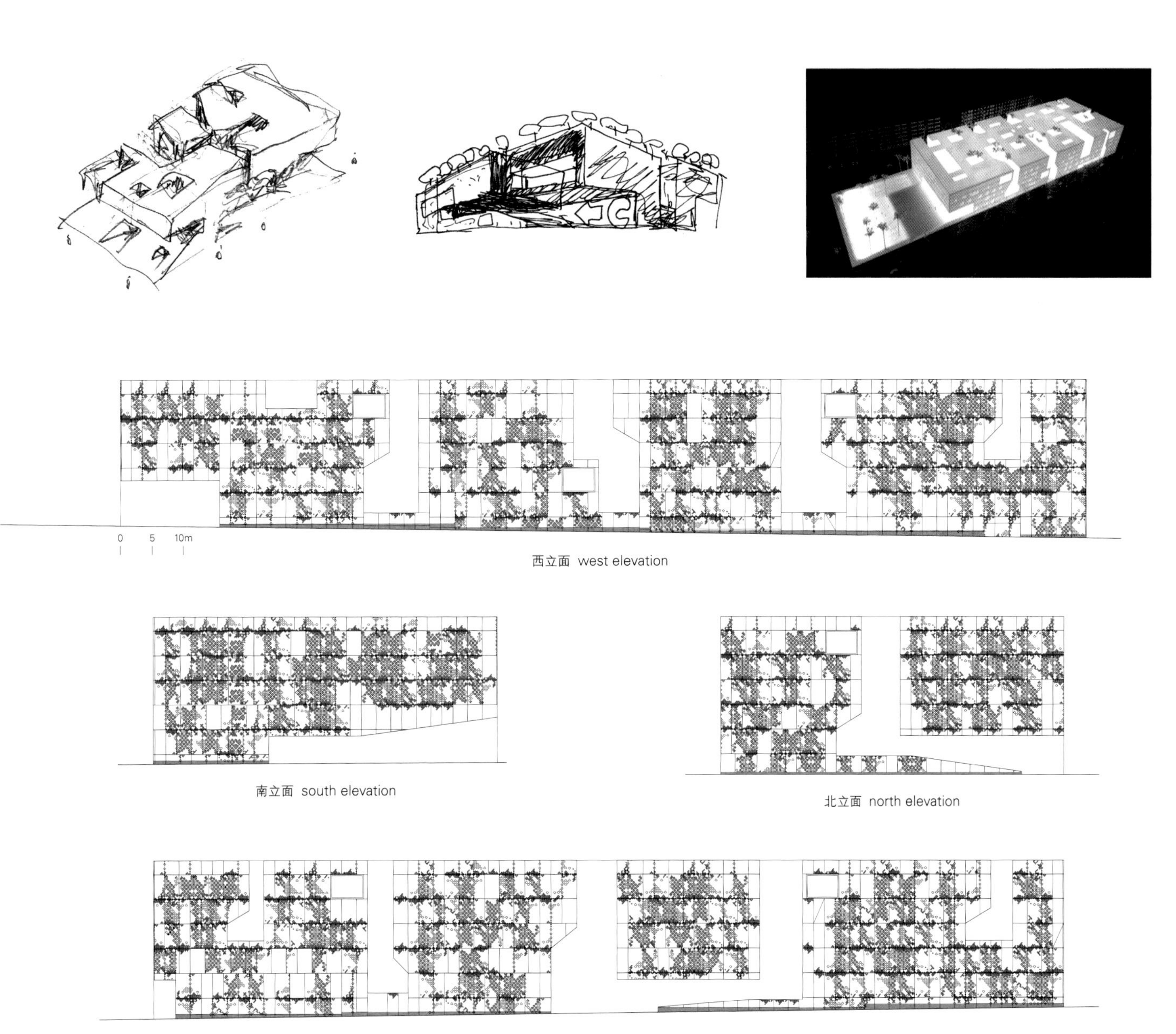

西立面 west elevation

南立面 south elevation

北立面 north elevation

东立面 east elevation

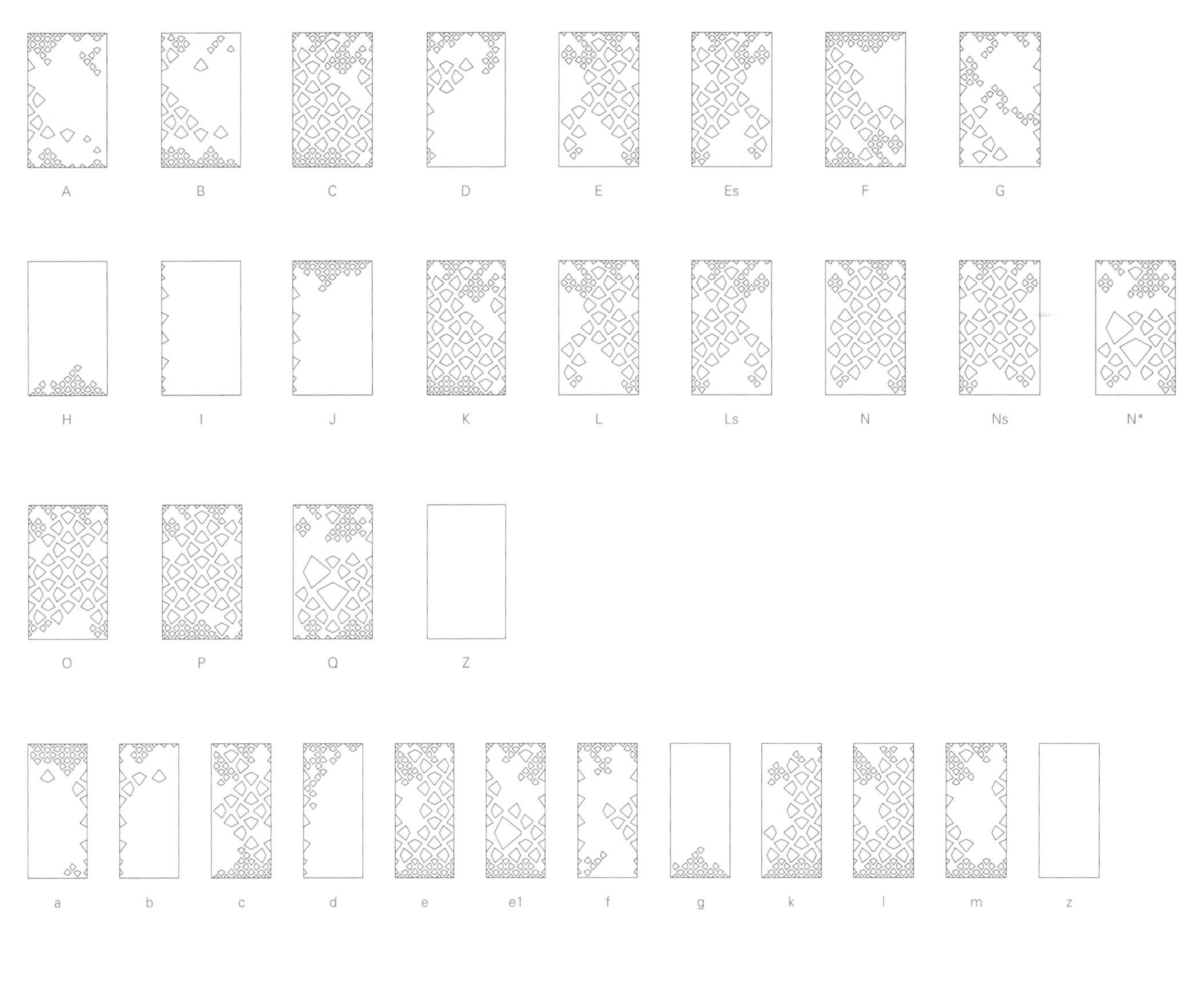

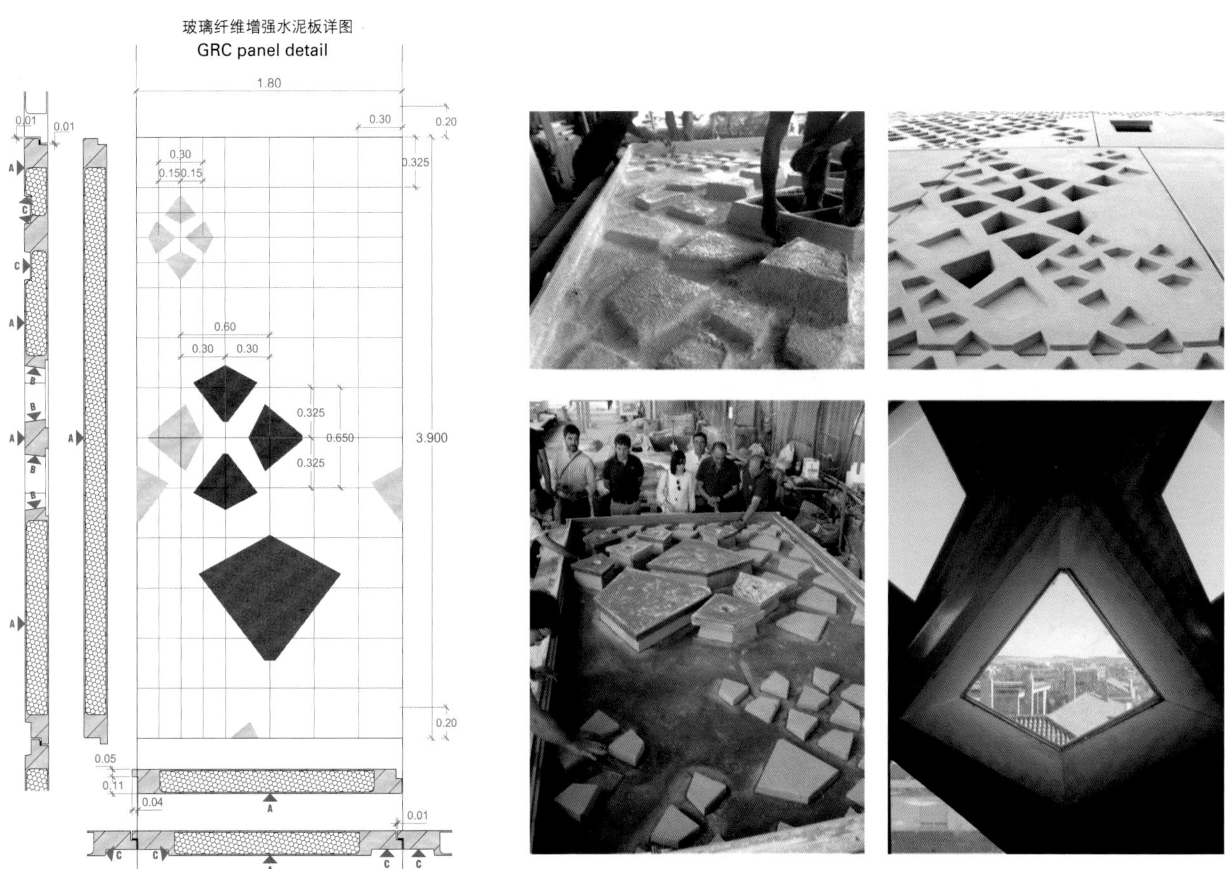

玻璃纤维增强水泥板详图
GRC panel detail

二层 first floor

一层 ground floor

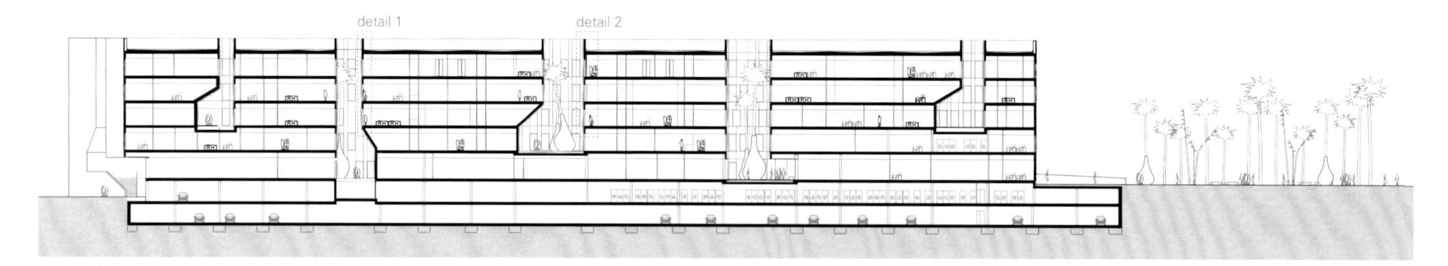

A–A' 剖面图 section A-A'

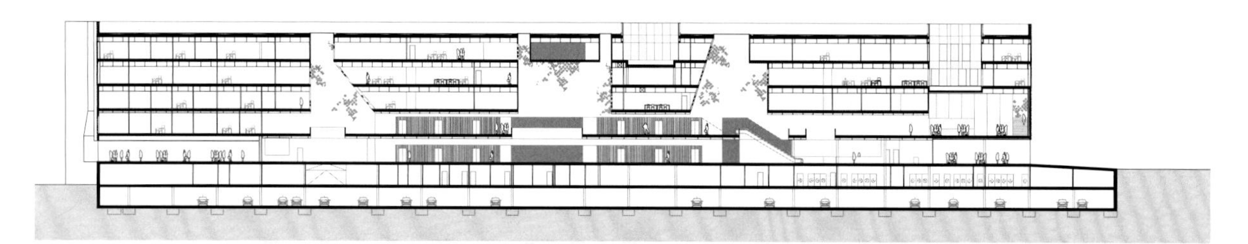

B–B' 剖面图 section B-B'

项目名称：Palace of Justice / 地点：Calle Isla Mallorca, Calle Isla Formentera, Calle Isla Gomera, Calle Cantabrico, Córdoba, Spain / 建筑师：Mecanoo Architecten + AYESA / 客户：Consejeria de Justicia e Interior (Junta de Andalucia) / 顾问：Consultants (energy efficiency, structural engineer, mechanical engineering, electrical engineer, lighting, fire safety, roof and facade): AYESA / 承包商：UTE ISOLUX CORSAN-COPCISA / 用途：courthouse with 26 courtrooms, a wedding room, a Forensic Institute, offices, a cafe, an archive, a prison and a parking garage / 建筑面积：48,000m² / 设计时间：2006 / 施工时间：2014—2017 / 摄影师：©Fernando Alda (courtesy of the architect)

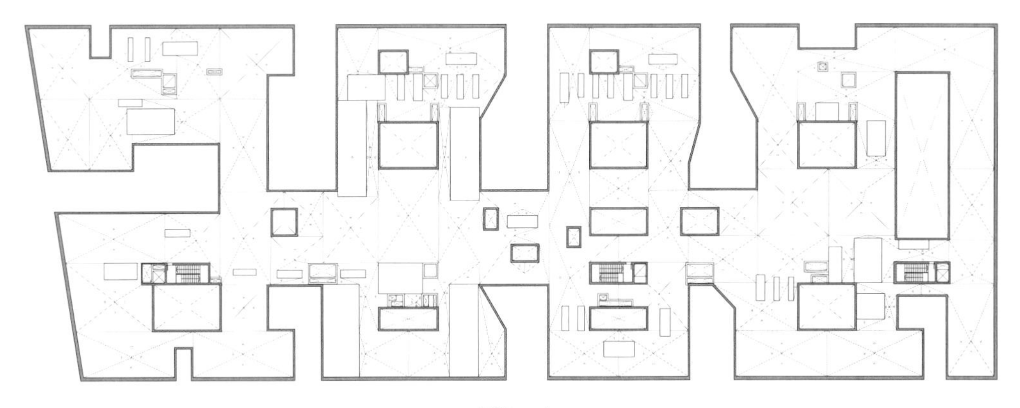

屋顶 roof

五层 fourth floor

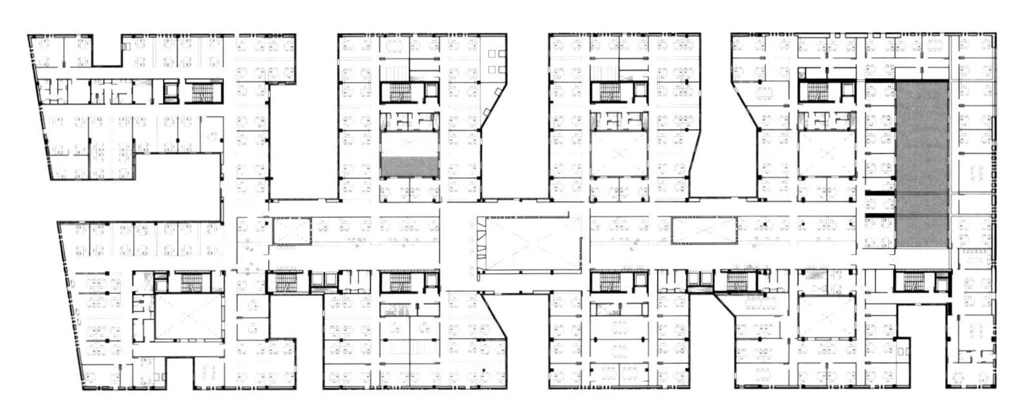

四层 third floor

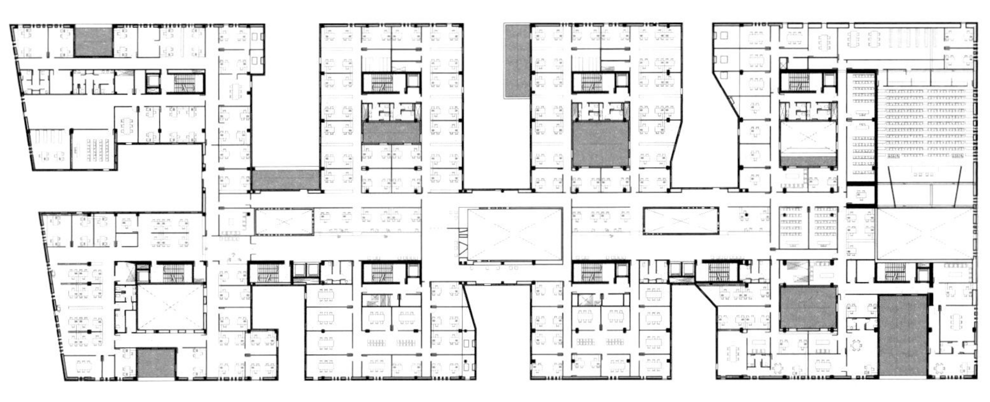

三层 second floor

SALIDA

The vertical fractures that are introduced in the building mass create patios, relating to the local courtyard typologies. These fractures provide natural light and ventilation in the central zones of the large building. One can say that the sustainability of the building is not achieved by expensive technological mechanisms but by an intelligent interpretation of the vernacular architecture.

The building is elevated two meters from the street level. This elevation gives a symbolic power to the building while solving problems of privacy and security created by the insertion of semi-open patios. Since the Palace is divided into several independent departments, the building requires several entrances that are positioned based on both internal and external hierarchies. Access to the building is either from the main entrance square or smaller open patios carefully placed along the other 3 sides of the building.

From the main entrance, the interior organization is easily recognizable. A central spine creates a circulation axis which connects to the various programs of the building. This space spans through several floors across the length of the building and articulates each department. It also links the public circulation with the exterior patios. The spine echoes the architectural language of the exterior massing, with multi-level day-lit voids creating sculptural atriums throughout the building.

The internal functions become more private higher up the building. At the level of the square, the courthouse features an open ground floor that contains the most public sections such as courtrooms, marriage registry, and restaurant. High-security offices are situated off the upper courtyards while the archives and jail cells are found below the ground level.

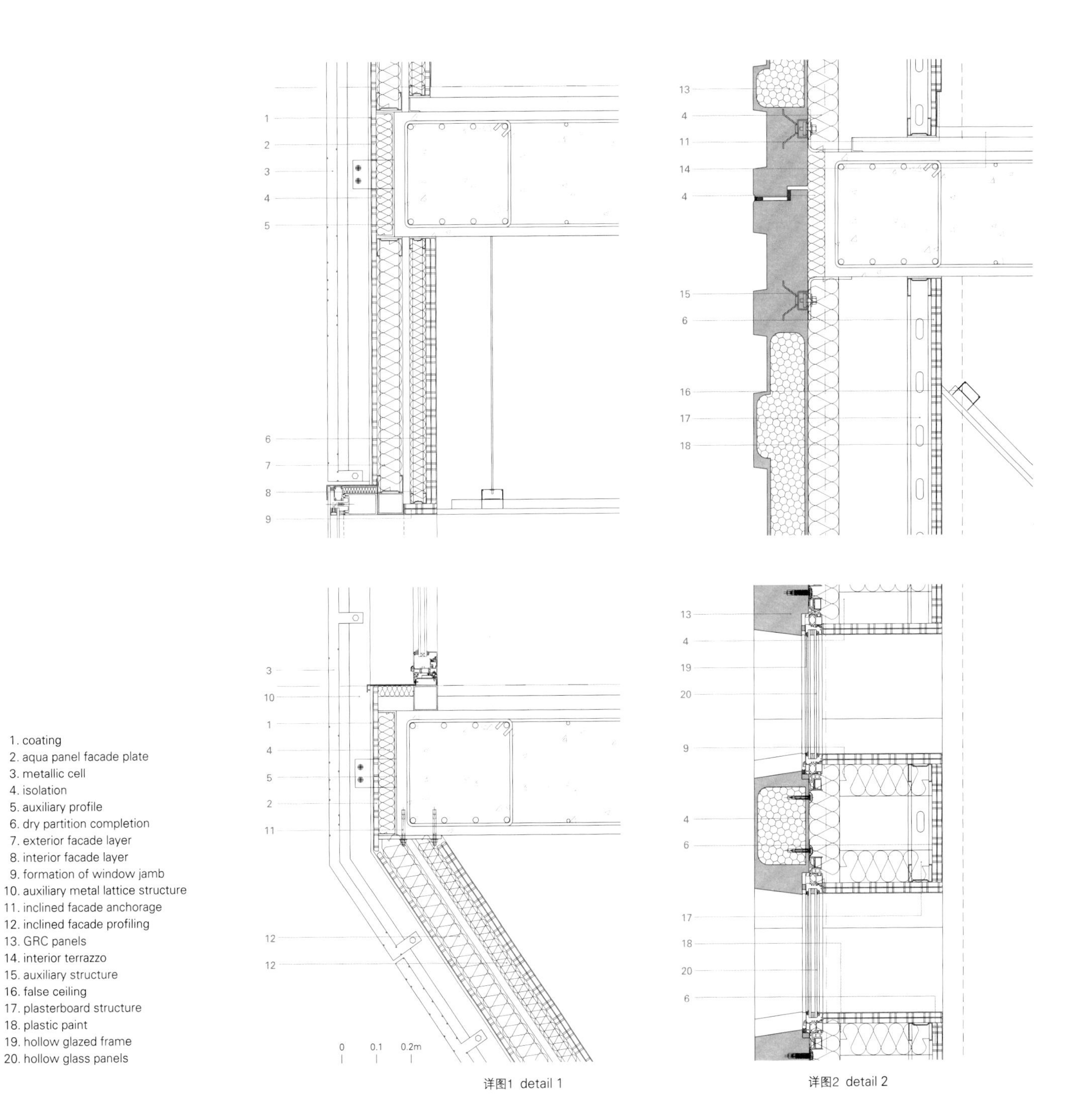

详图1 detail 1

详图2 detail 2

巴黎法院
Paris Courthouse

Renzo Piano Building Workshop

自中世纪以来，巴黎的法院机构就开始在西提岛上圣礼拜堂附近的著名建筑之中安家落户。然而这些年来，空间短缺现象越来越严重，许多办公室已经转移到巴黎多个其他地点。因此法国政府发起了巴黎法院的建筑设计竞赛，提议为巴黎法院建造两栋独立的大楼，每栋楼都设有法庭和办公室。然而，伦佐·皮亚诺建筑工作室的主要目标是将司法机构的法庭、600个办公室和其他公共区域（如图书馆）统一安置在一座大型建筑内，以期通过其建筑规模和地位，使其成为该地区复兴和重新发展的开端。

这座建筑从位于城市环路和马丁·路德·金公园之间一个L形场地上拔地而起，建筑的轴线与马丁·路德·金公园的南北轴线互相平行，而法院的建造也成为克里希–巴蒂尼诺利斯城市开发区发展的决定性因素。大楼东西两侧立面的细玻璃"框架"延伸到了建筑之外，朝向蒙马特区和埃菲尔铁塔。南北立面较窄，分别面向巴黎市中心或者克里希和瓦尔里恩山。

该建筑底部是一个六到九层高的基座，上边依次堆叠了三个体块。该塔状建筑物上的体块随着高度升高向内缩进，形成了一个独特的阶梯状轮廓，矗立于巴黎的上空，在传统建筑的映衬下显得极有辨识度。

游客和工作人员可经由建筑地面层的前庭进入公共大厅，并在此受到接待和指引。"大中庭"位于基座建筑中心，具有象征性意义，是公共大厅的代表和标志。这个长方形中庭与基座建筑等高，中庭中的细长钢柱非常引人注目，大量阳光从天窗和玻璃立面倾泻而入。一条160m长的走廊从北向南纵向穿过一层，连接着另外两个较小的中庭。

公共大厅可以通向所有公共与服务区域，包括会议室和公共自助餐厅，以及90个法庭。所有房间都装有镶木地板和蒸山毛榉木镶板，几乎所有的房间都能享受到穿墙而入的日光。特别是人们在室外可以看到法庭后面的会议室和审议室。

九层有一个种有植被的7000m²露台和一个员工餐厅，供法院工作人员散步、沉思以及私下交流互动之用。建筑的20层和30层建有花园，塔楼上的浮式体块在建筑的高层留出了建造花园的空间，因此马丁·路德·金公园得以"延伸"到建筑中，创造出了一栋真正意义上被植物所覆盖的摩天大楼。

东西立面有两个垂直"脊柱"，连接着三个浮式塔状体块。这两个"脊柱"由光伏板组成，内部分别设有外部电梯和阳台。光伏板在阳光的照射下呈现出明亮而充满活力的光泽，也表达出设计团队将这项技术在其他公共建筑中进行推广的愿望。

大楼的结构呈现横平竖直的外形，非常坚固，同时该建筑还具有灵活性，能够适应未来的需求，在未来司法系统运行方式发生变化时也能够灵活应对。

Since the Middle Ages, Parisian justice has been dispensed from the famous building that surrounds the Sainte-Chapelle on the Île de la Cité. However, over the years, an increasing shortage of space has transferred many of its offices to a multitude of locations across the city. When the competition was first launched, the French government suggested the new Paris law courts to have two separate buildings, each accommodating courtrooms and offices. However, the key aim of Renzo Piano Building Workshop was to reunite judicial institution's courtrooms, 600 offices and additional communal areas like library in one large building: capable, through its size and status, of becoming a starting point for the rehabilitation and redevelopment of the neighborhood.

The building rises out of an L-shaped site, between the city ring road and the Martin Luther King park. The building's axis is aligned with the north-south diagonal of the adjacent park, which anchors the Clichy–Batignolles urban development zone. The office facades to the east and west extend beyond the building to Montmartre and the Eiffel Tower, through fine glass "frames". The north and south facades, which are narrower, offer views of central Paris or Clichy and Mont-Valérien. The building is made up of a five to eight stories high Pedestal that integrates the lower part of the Tower of three superimposed parallelepipeds, levitating one above the other. The blocks are set back as the tower rises, creating a distinctive step-like profile, distinguishable in the conventional Paris's skyline.

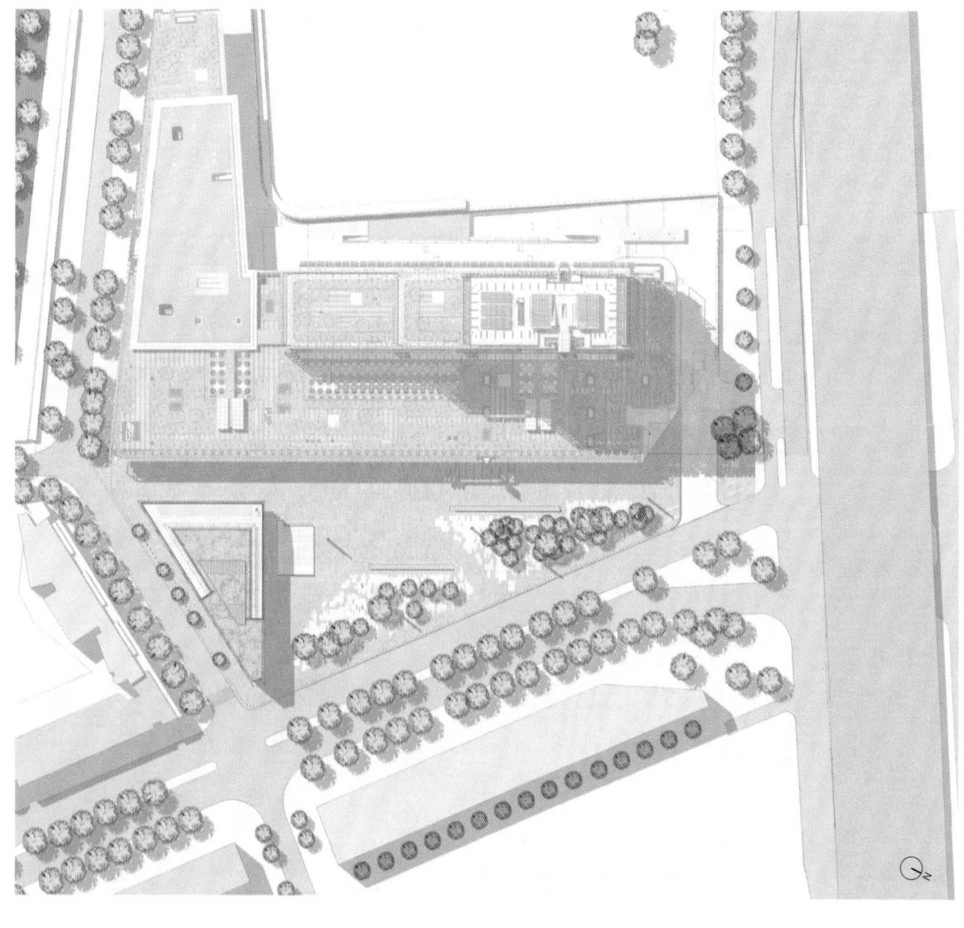

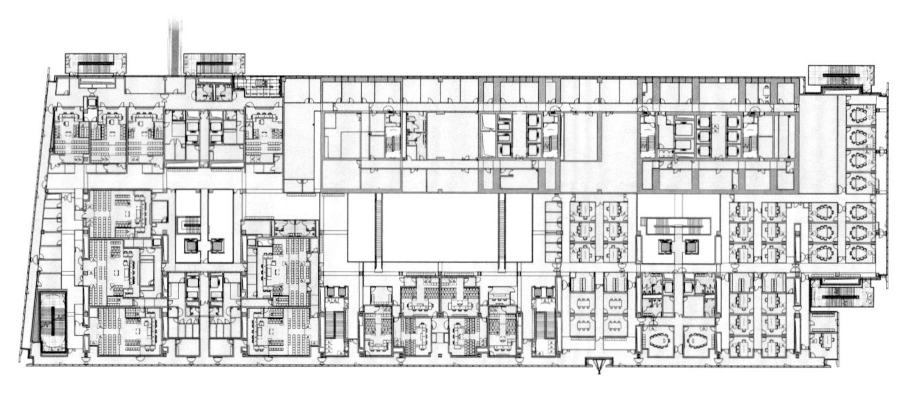

七层 sixth floor

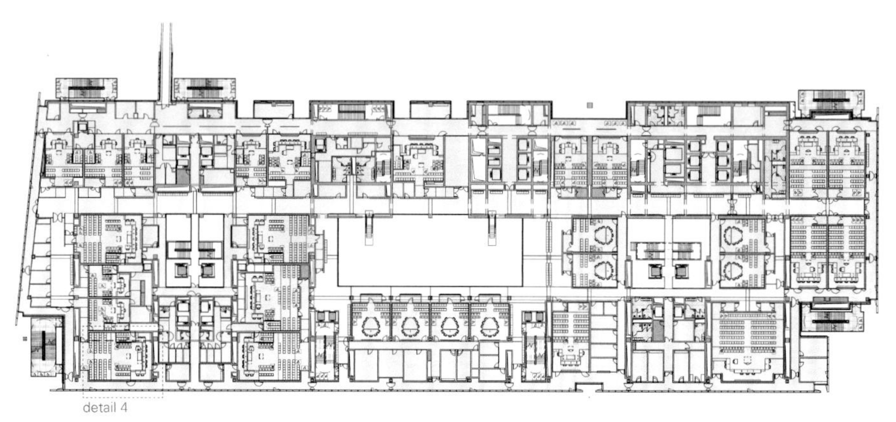

五层 fourth floor

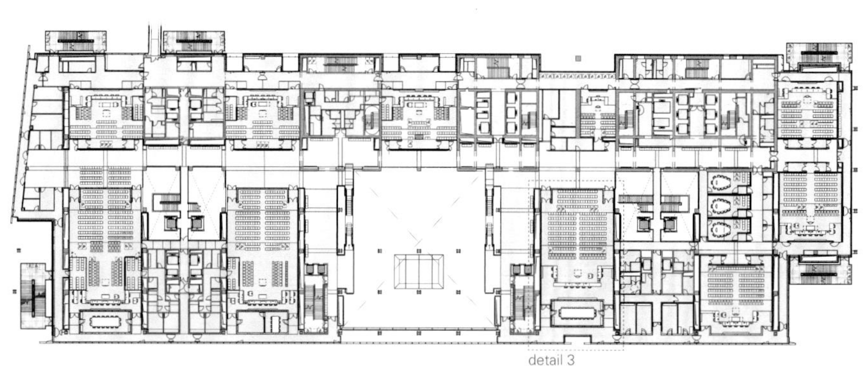

三层 second floor

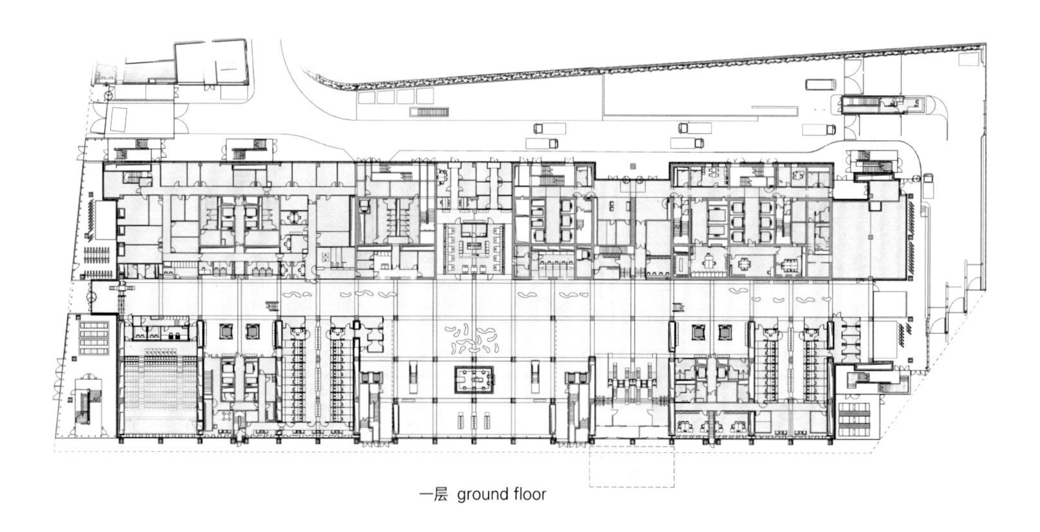

一层 ground floor

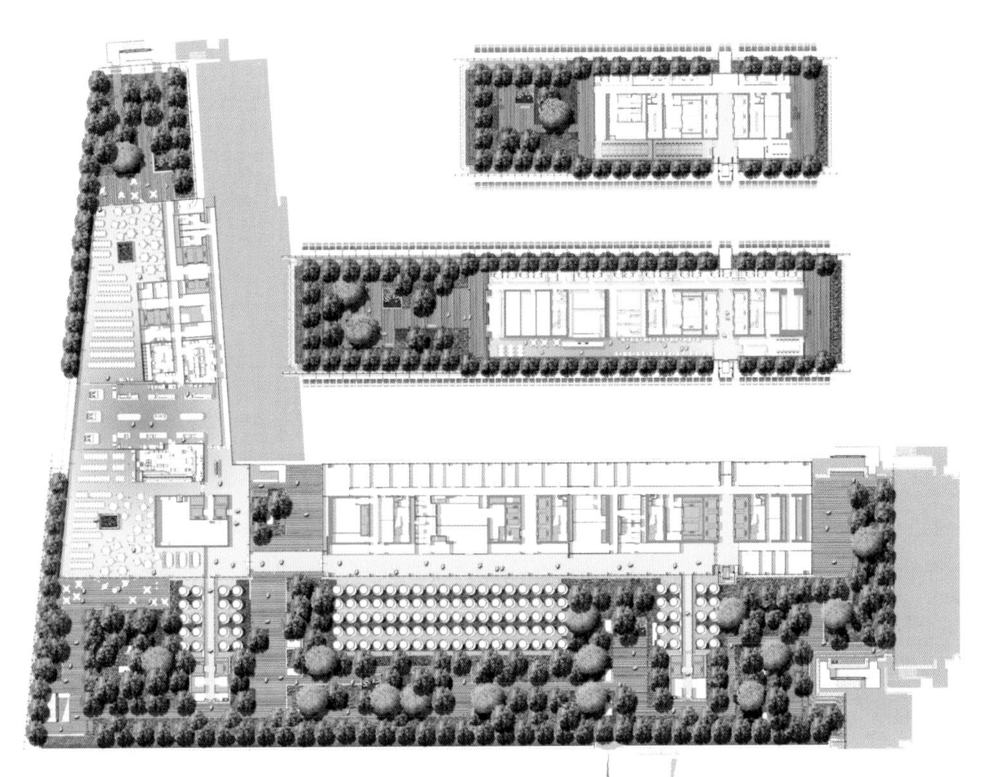

露台景观 terrace landscape

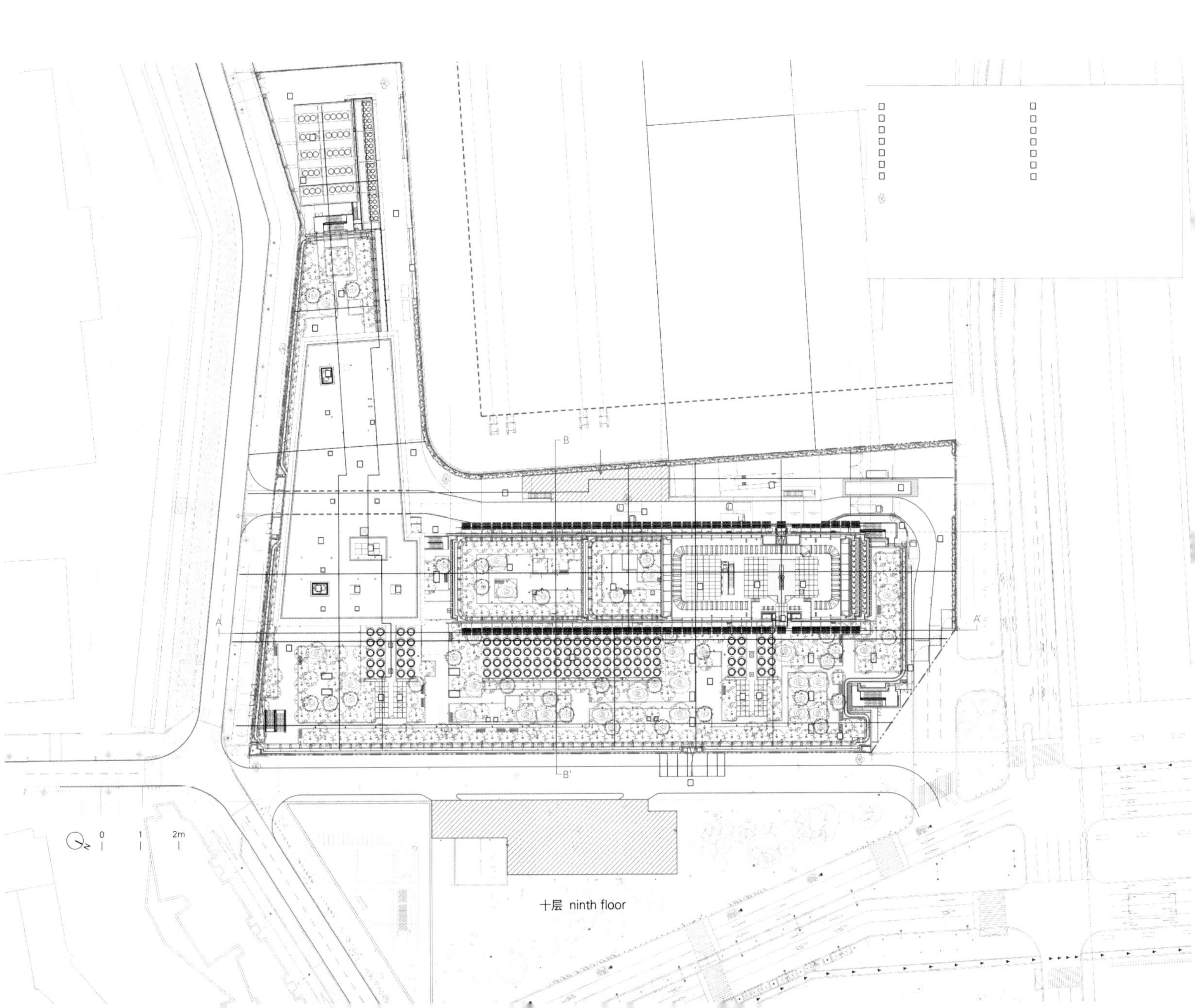

十层 ninth floor

Entered from the ground-level forecourt into the Public Lobby, the flux of visitors and employees is greeted and directed. The Public Lobby is represented by a "Great Atrium", set symbolically right at the heart of the Pedestal. The rectangular atrium of full height of the Pedestal, is notable for its slender steel columns and the amount of natural light that enters through its skylights – "the Marilyns" – and the glazed facade. A 160-meter-long corridor crosses the ground floor longitudinally from north to south, additionally connecting to the two smaller atriums.
The lobby leads to all public posts and services including a meeting room and public cafeteria, as well as the 90 courtrooms. Fitted with parquet and steamed-beech-wood paneling, almost all the rooms benefit from daylight that filters through the facades. Especially, the council chamber and deliberation rooms behind the courtrooms are visible from the outside.
The eighth floor is home to a 7,000m² planted terrace, as well as the staff restaurant, dedicated for walking, reflecting, and informal interaction between Courthouse staff members. On the 19th and 29th floors, the Tower's floating blocks create space for two further raised gardens, allowing the Martin Luther King park to "extend" into the building and creating a genuinely plant-covered skyscraper.

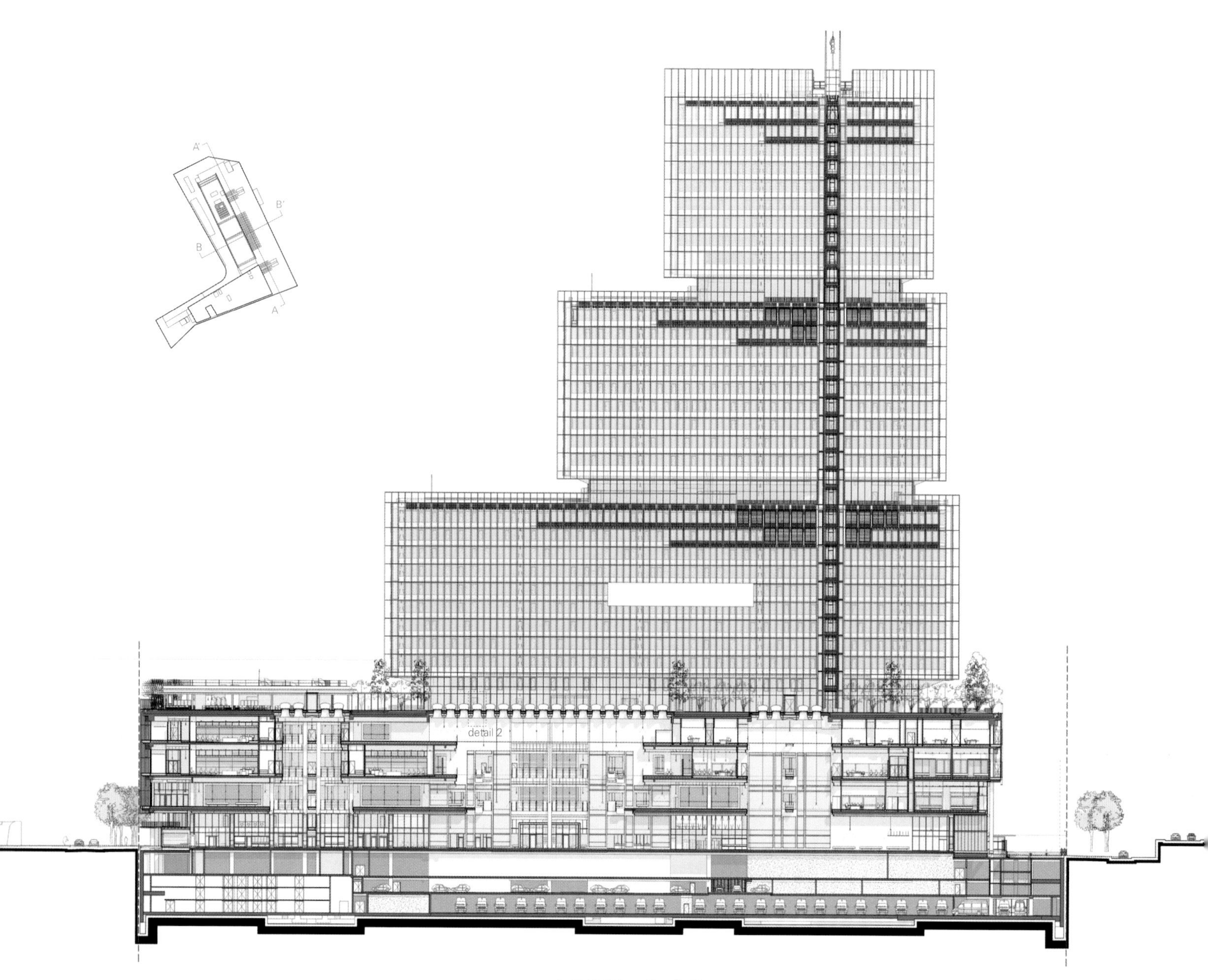

A–A' 剖面图 section A-A'

Two vertical "spines" on the east- and west-facing facades that link the three floating Tower blocks, are made up of photovoltaic panels, and contain external elevator and balconies respectively. Photovoltaic panels create vibrancy accentuated by the light that reflects off them and demonstrate a desire to move toward the usage of alternative energy in public buildings.
The building's structure, robust and orthogonal, ensures a flexibility that, over the long term, will be able to accommodate future requirements, including any changes in the way the justice system operates.

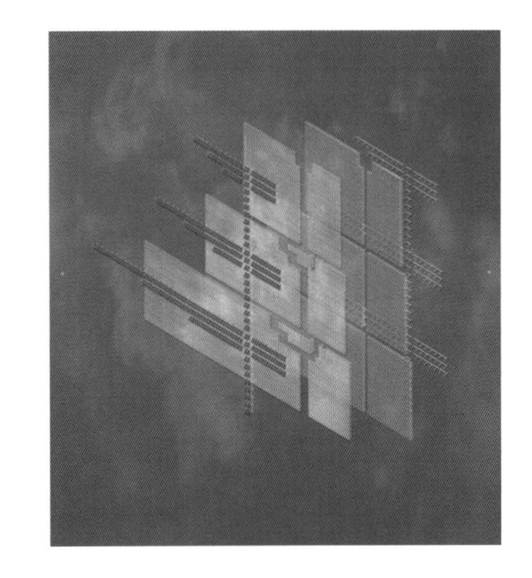

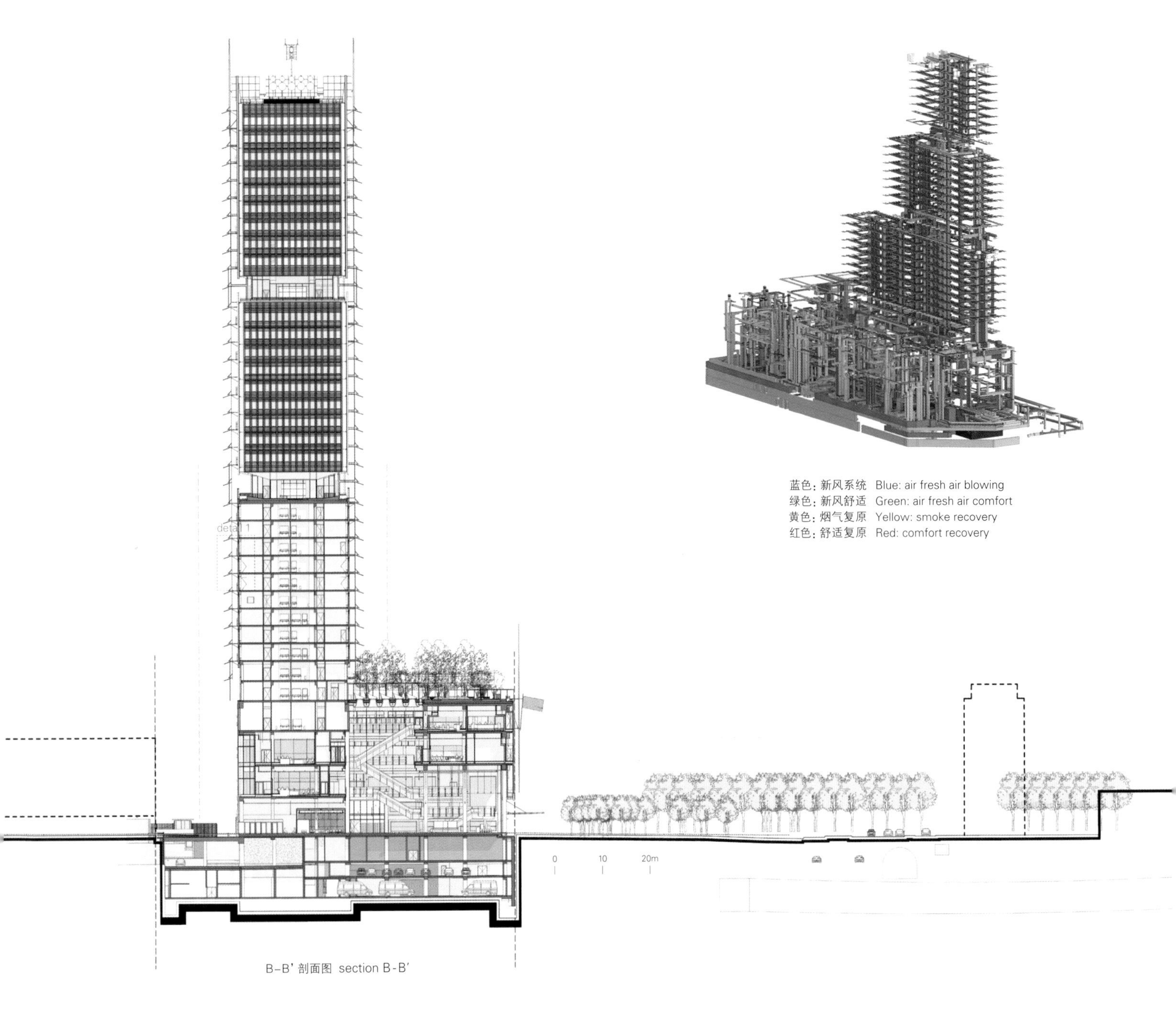

B-B' 剖面图 section B-B'

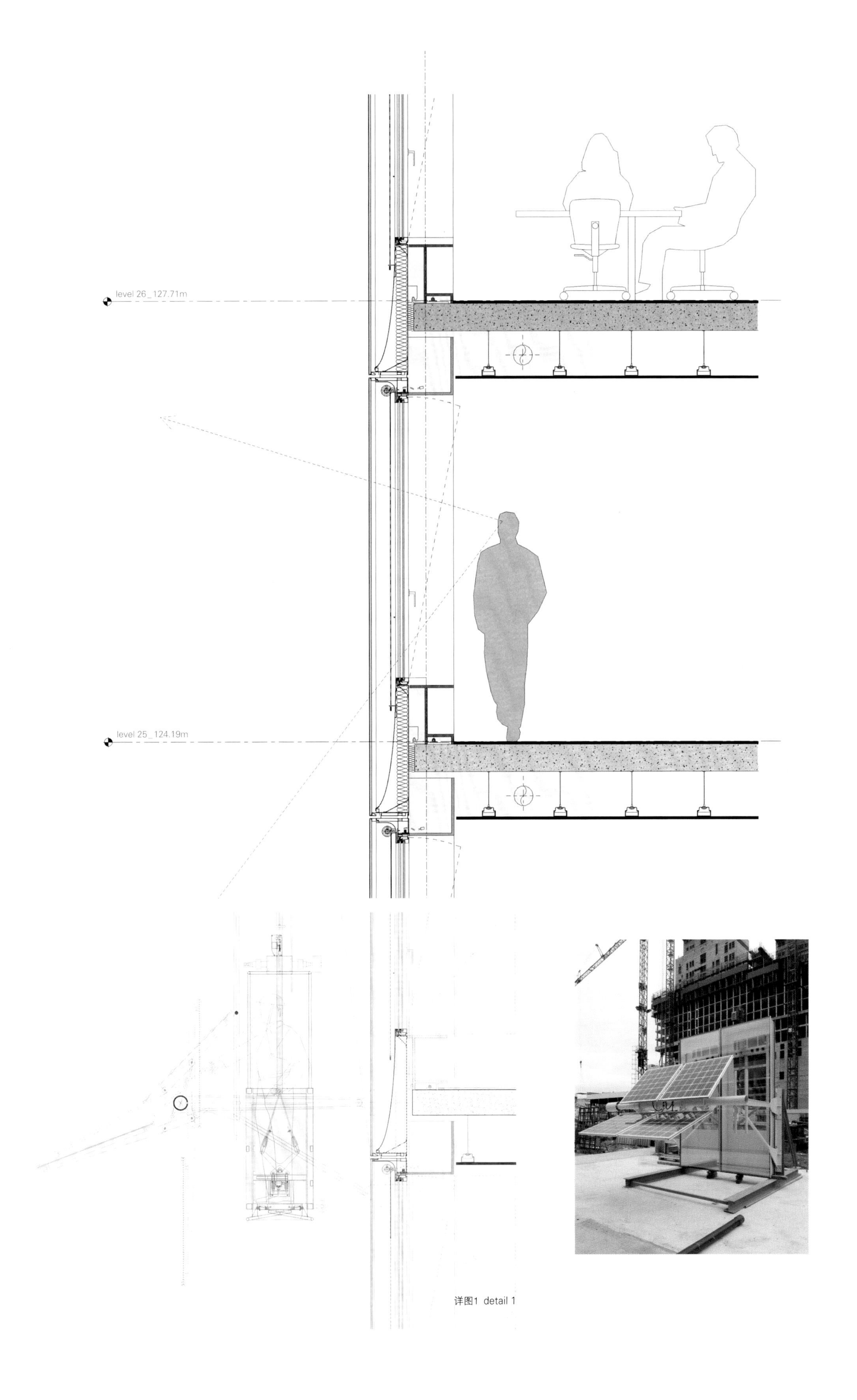

详图1 detail 1

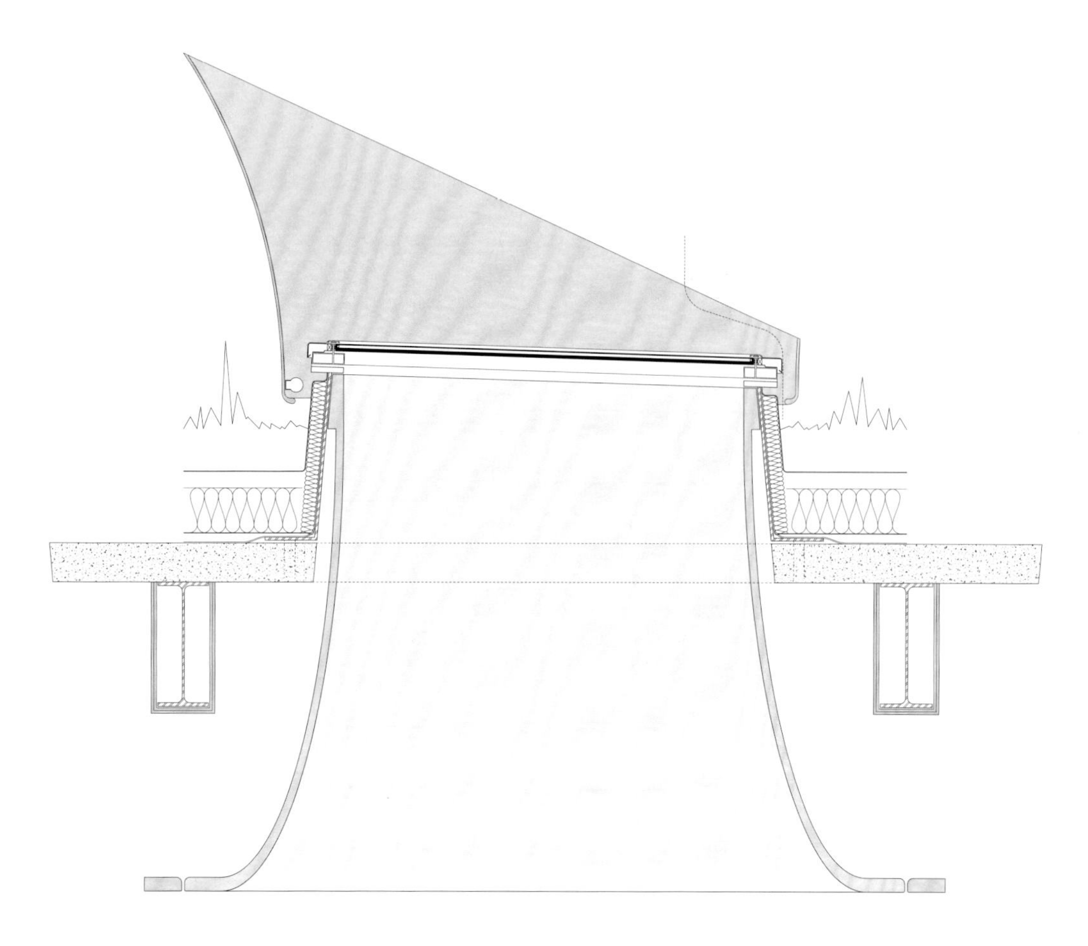

详图2 detail 2

6
4
2
2
1

项目名称：Paris Courthouse / 地点：Zac de Clichy-Batignolles, Porte de Clichy, Paris 17th arrondissement, France / 建筑师：Renzo Piano Building Workshop / 竞赛设计团队：B. Plattner-partner in charge, J.B.Mothes, D.Rat with B.Akkerhuis, M.Angelozzi, L.Bot, N.Byrelid, S.Cloarec, S.Crabot, J.Franco, F.Garrigues-Cortina, S.Giorgio-Marrano, L.Le Roy, J.Moolhuijzen-partner, P.Pires da Fonte, B.Schelstraete, O.Aubert, C.Colson, Y.Kyrkos, Y.Chaplain-models / 竞赛顾问：SETEC Bâtiment; Berim-MEP; SETEC TPI-structure; Eléments Ingénieries-sustainability; RFR-façades; Movveo-vertical transportation; Majorelle-interior design, space planning / 设计开发团队：B.Plattner-partner in charge, S.Giorgio-Marrano, J.B.Mothes, D.Rat-associates in charge, G.Chung, S.Cloarec, B.Granet, A.Greig, C.Guézet, A.Karcher, M.Sismondini with N.Aureau, G.Avventi, A.Belvedere-partner, A.Bercier, F.Bolle, L.Bot, A.Boucsein, N.Byrelid, J.Chevreux, S.Crabot, J.Franco, N.Grawitz, B.Guimaraes, V.Houeiss, N.Maes, J.Sobreiro, S.Stevens, M.van der Staay and T.Heltzel,

M.Matthews, O.Aubert, C.Colson, Y.Kyrkos, Y.Chaplain-models / 设计开发顾问：SETEC Bâtiment, Berim-MEP; SETEC TPI-structure; ELAN-sustainability; RFR-façades; Movveo-vertical transportation; Lamoureux-acoustics; M.Harlé/J.Cottencin-signage; Cosil Peutz-lighting; C.Guinaudeau, AIA Ingénierie-planting; Majorelle-interior design, space planning; Ecotec-specification consultant; Studio Akkerhuis-consulting architect) / 客户：Etablissement Public du Palais de Justice de Paris + Bouygues Bâtiment / 用地面积：17,500m² / G总建筑面积：142,000m² of which: IGH 53,000m², basement 37,000m² and bastion 20,000m² / 竞赛时间：2010—2011 / 设计开发时间：2012—2017 / 摄影师：©Sergio Grazia (courtesy of the architect)-p.156~157, p.159, p.164; ©Michel Denance (courtesy of the architect)-p.167, p.168~169, p.171; ©Francesca Avanzinelli (courtesy of the architect)-p.165, p.166; ©Florent Michel (courtesy of the architect)-p.155

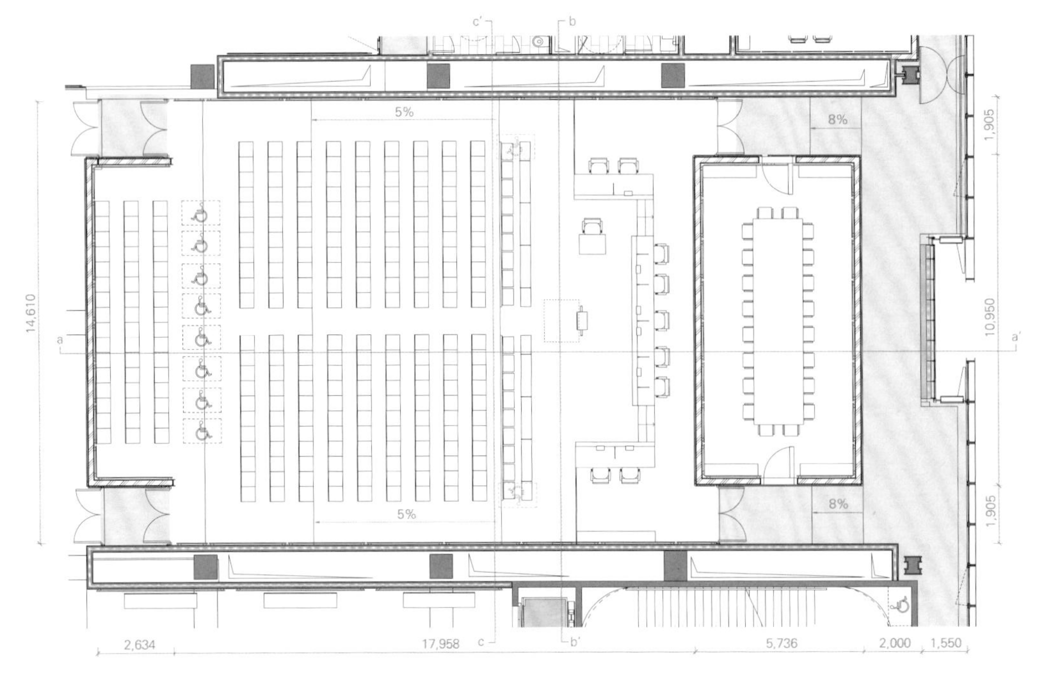

详图3 detail 3

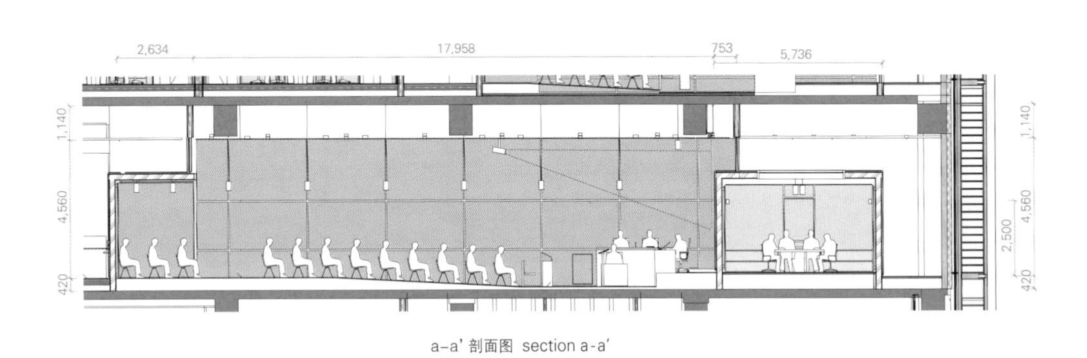

a–a' 剖面图 section a-a'

b–b' 剖面图 section b-b'

c–c' 剖面图 section c-c'

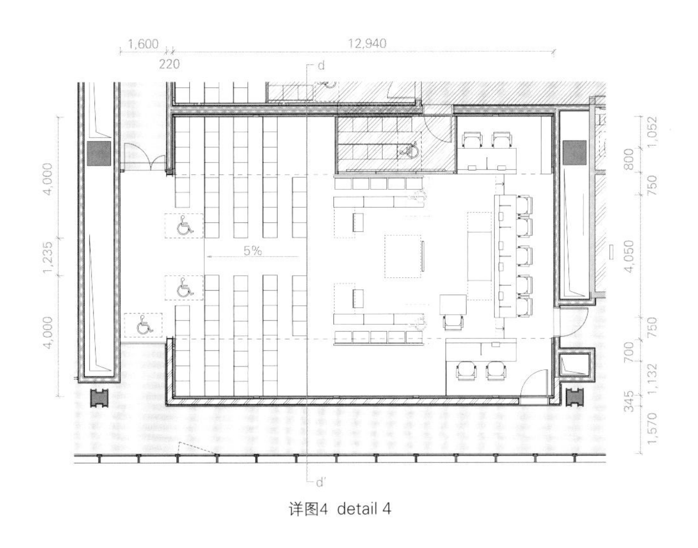
详图4 detail 4

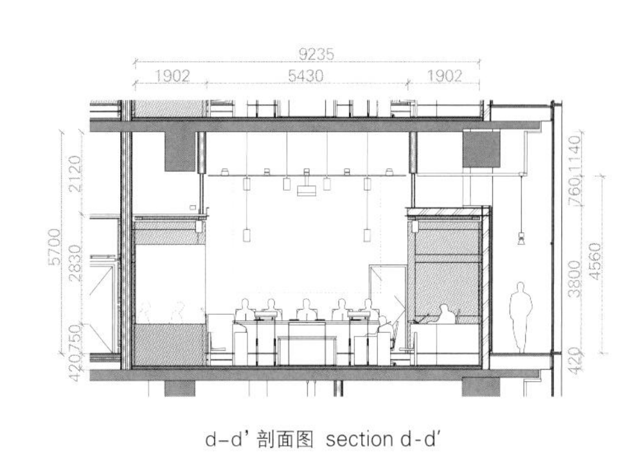
d–d' 剖面图 section d-d'

本项目要求设计师根据现场的地形和监管环境，对建筑的城市和建筑特征进行定制设计。

在本项目的开发过程中，我们必须解决两个问题：1、街道与新法院的关系以及如何通过前庭（一个公共机构空间）将街道与法院连接起来，以及2、建筑形象。富瓦法院所在地并非典型的城市区域，因此经典的正面建筑视角不能恰当地解决以上两个问题。

因此该建筑采用了圆形的建筑形态，在不偏向任何一个合成轴的情况下与场地融为一体。建筑与环境有所有的元素相交互，同时又不受周围的地理、景观或是人造结构的影响。该项目呈现了一种强有力的、可见的城市建筑形态，吸引了人们的注意力，并在不遮蔽它们的情况下构成了场地环境中多种多样的组成部分。

富瓦法院建在山坡之上，建筑与斜坡的相对位置可将其对地面原有水平的影响降至最低，从而使建筑能够与周围环境相融合，不会危害环境。这座法院大楼的规模和体量完全不会影响和干扰附近的景观，而在法院中人们总是能够将周围风光尽收眼底。法院广场坐落在朝向市中心的缓坡上。设计师充分利用地形特征，设计开拓出一条清晰可辨的路线通向法院。这一主要的室外空间与周围的景观相互协调，将新法院与苏德大道连接起来。缓坡广场直接向上延伸便是法院大厅，又称为"salle des pas perdues"，它将圆形庭院一分为二，成为各种公共活动的中心。

圆柱形天井为建筑提供了大量的自然光线，这些光线穿过广场上的树木射入天井。它们还对"salle des pas perdues"的线性交叉空间进行了构造和重新划分。圆形设计为建筑带来了极高的功能效率；设计师在建筑中设计了两条中央楼梯，将使用者的步行距离缩到最短，使人们既可以集中又可分散地进入建筑的四个管辖区。任何地点，无论是地方法官入口或是"salle des pas perdues"入口，都可以是建筑内部路线的开端。最后，虽然内部空间易于改动，但法院的建筑形状将保持不变。

法庭集中安置在大厅的一侧，另一侧则连通到公共接待处和商务法庭，人们也可以从独立的建筑主入口进入商务法庭。

法庭位于阶梯大厅之间，涵盖两个级别的司法管辖区。飘窗用丝网框架做了处理，可以俯瞰室外的树木，丝网框架的设计可以确保室内光线环境均匀且无眩光，同时也保护了房间隐私，防止人们从外部窥视。轻木制成的地板和装饰天花板以及家具设备，共同营造了一个清新明亮的氛围，为空间带来了凝聚力。

其余室外空间的设计以整合为优先目标。与周围环境相融合是项目驱动目标的延伸，即保持完整的建筑连贯性、限制环境足迹、最小化土方工程以及重新栽种所有屋顶区域的植被。

富瓦法院

Palace of Justice in Foix

Philippe Gazeau

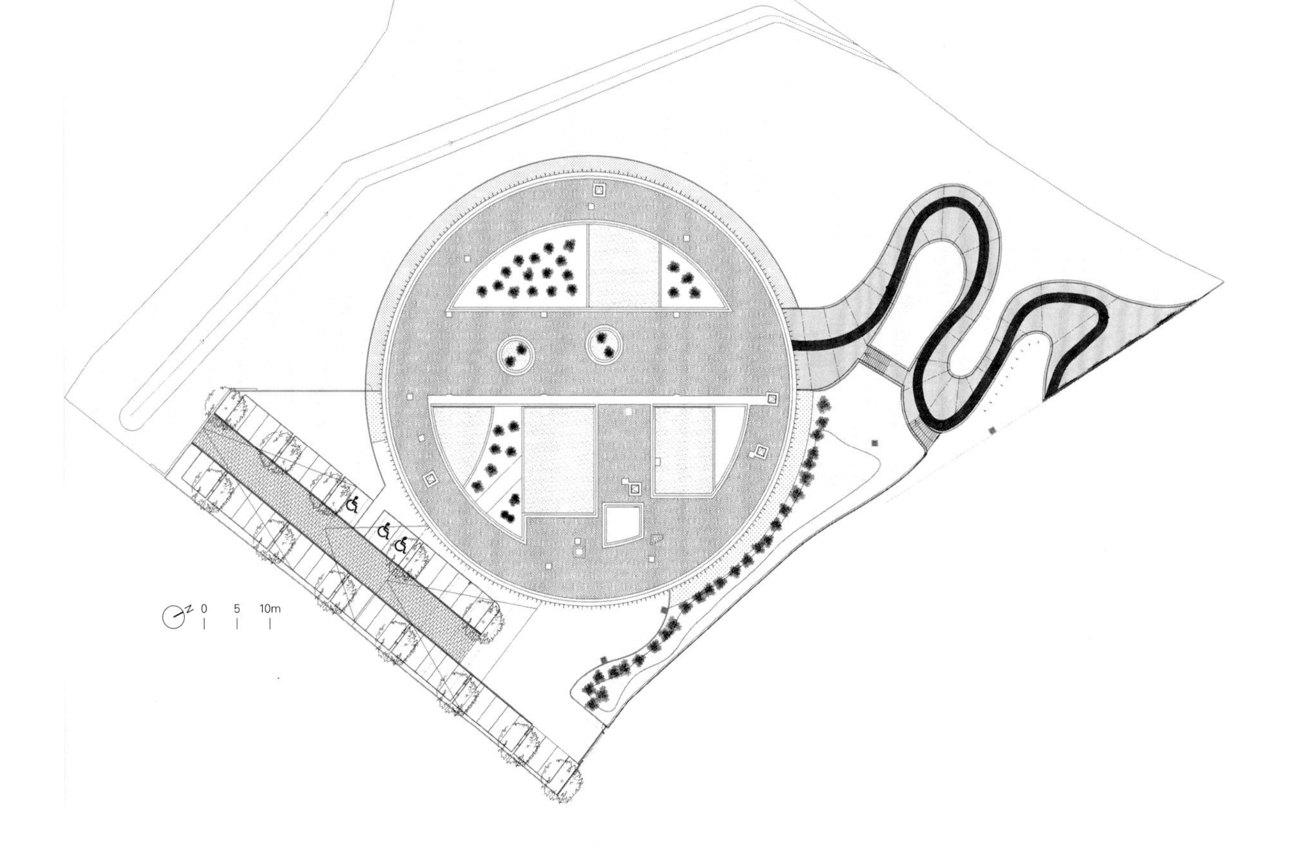

For this project, the topographic and regulatory environment of the site demanded an urban and architectural reflection for a structure customized to the context.
Two ideas had to be addressed in the development of this project: 1. The relationship between the street and the new Palace of Justice and the articulation thereof through the forecourt (an institutional public space), and 2. The architectural image. A classic frontal perspective could not properly cover both ideas in this atypical urban context.
That is why the project adopts a circular configuration, allowing it to fit into the site without favoring one composition axis over the other. It interacts with all elements in its environment, regardless of geography, landscape, or manmade structures. The project stands as a strong and visible architectural and urban shape that draws attention to and structures the heterogeneous components of the site without overshadowing them.
The position of the structure relative to the slope allows for minimal impact on the preexisting level of the ground, which enabled the project to be integrated with its surroundings without causing trauma to the environment. The volume of the courthouse never blocks the nearby landscape, always open to the different panoramas of the area. The plaza sits on a gentle, rolling slope oriented towards the town center. It makes full use of the topographic characteristics to create an identifiable and clear access route to the legal institution. This major outdoor space synergizes with the surrounding landscape and helps to bind the new courthouse to Boulevard du Sud. The courthouse hall, the "salle des pas perdues", is a direct extension of the gently-sloping plaza. It bisects the circular courtyard and becomes the heart and central element where various public activities may be organized.
The cylindrical patios provide generous natural light filtered

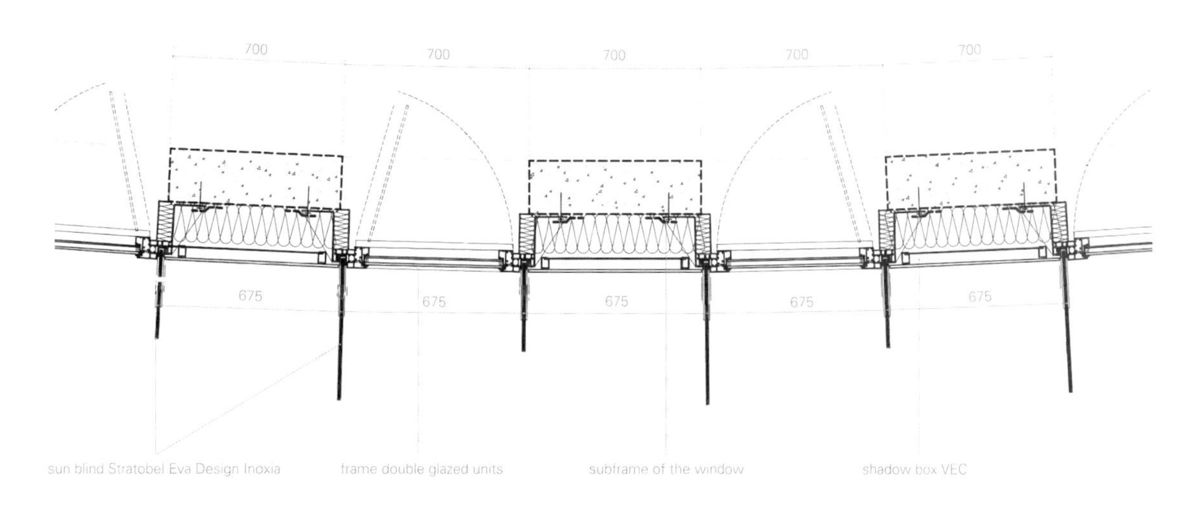
700
700
700
700
700
675
675
675
675
675
sun blind Stratobel Eva Design Inoxia
frame double glazed units
subframe of the window
shadow box VEC

详图1 detail 1

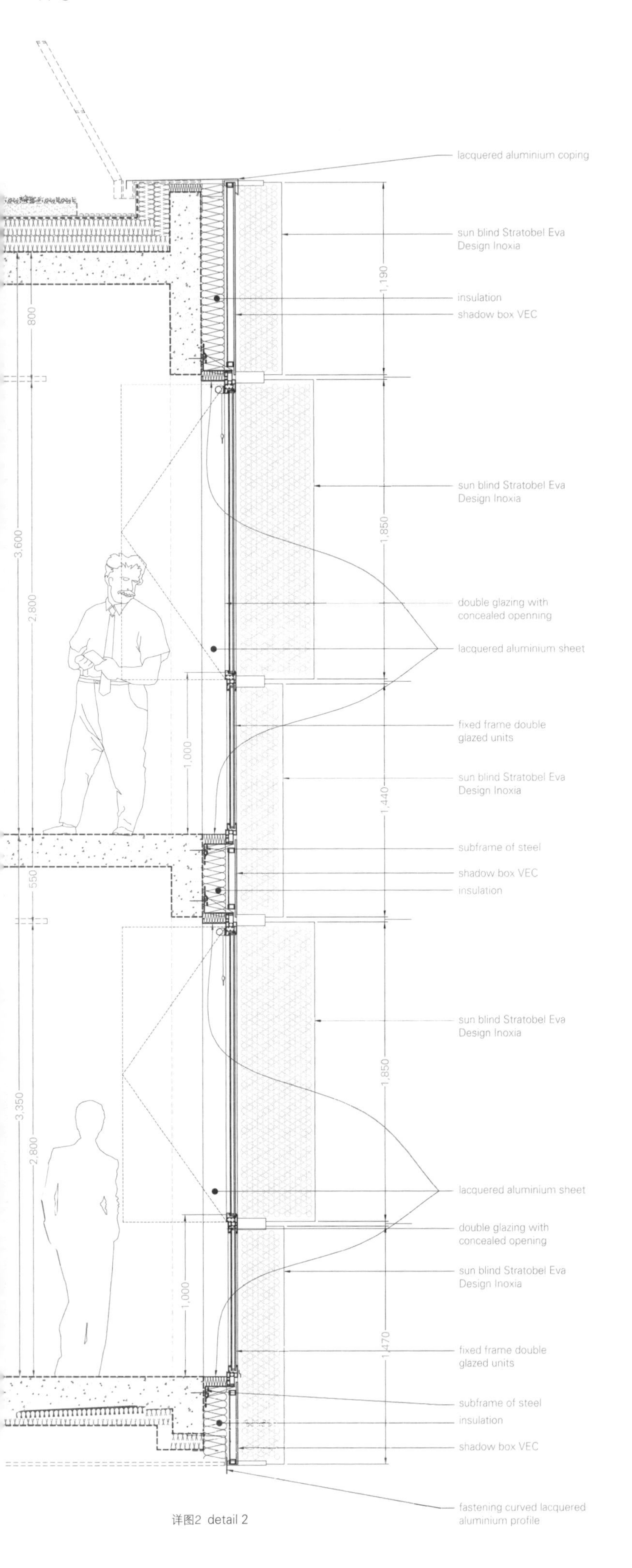
lacquered aluminium coping
sun blind Stratobel Eva Design Inoxia
insulation
shadow box VEC
sun blind Stratobel Eva Design Inoxia
double glazing with concealed openning
lacquered aluminium sheet
fixed frame double glazed units
sun blind Stratobel Eva Design Inoxia
subframe of steel
shadow box VEC
insulation
sun blind Stratobel Eva Design Inoxia
lacquered aluminium sheet
double glazing with concealed opening
sun blind Stratobel Eva Design Inoxia
fixed frame double glazed units
subframe of steel
insulation
shadow box VEC
fastening curved lacquered aluminium profile
1,190
800
1,850
3,600
2,800
1,000
1,440
550
1,850
3,350
2,800
1,000
1,470

详图2 detail 2

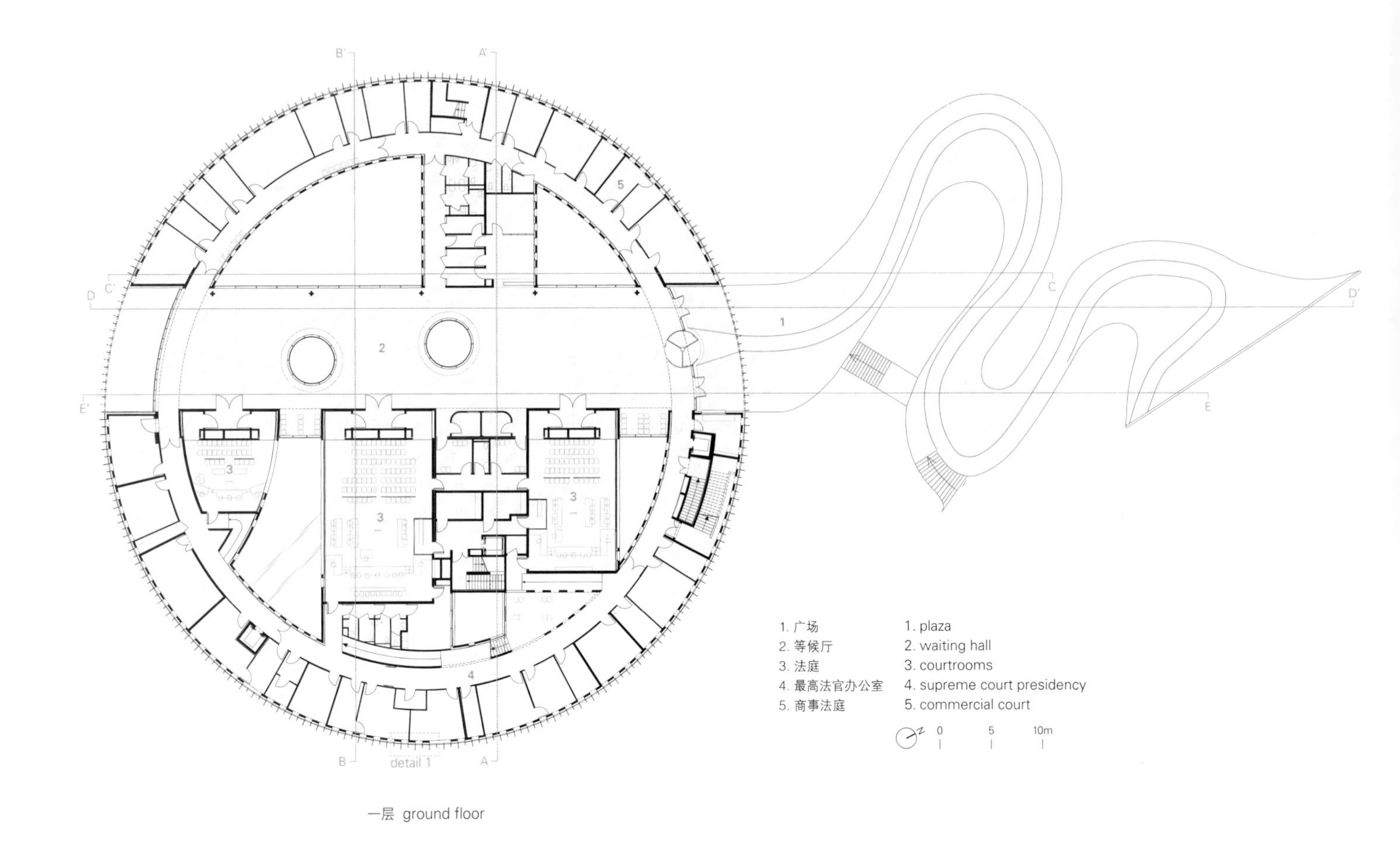

一层 ground floor

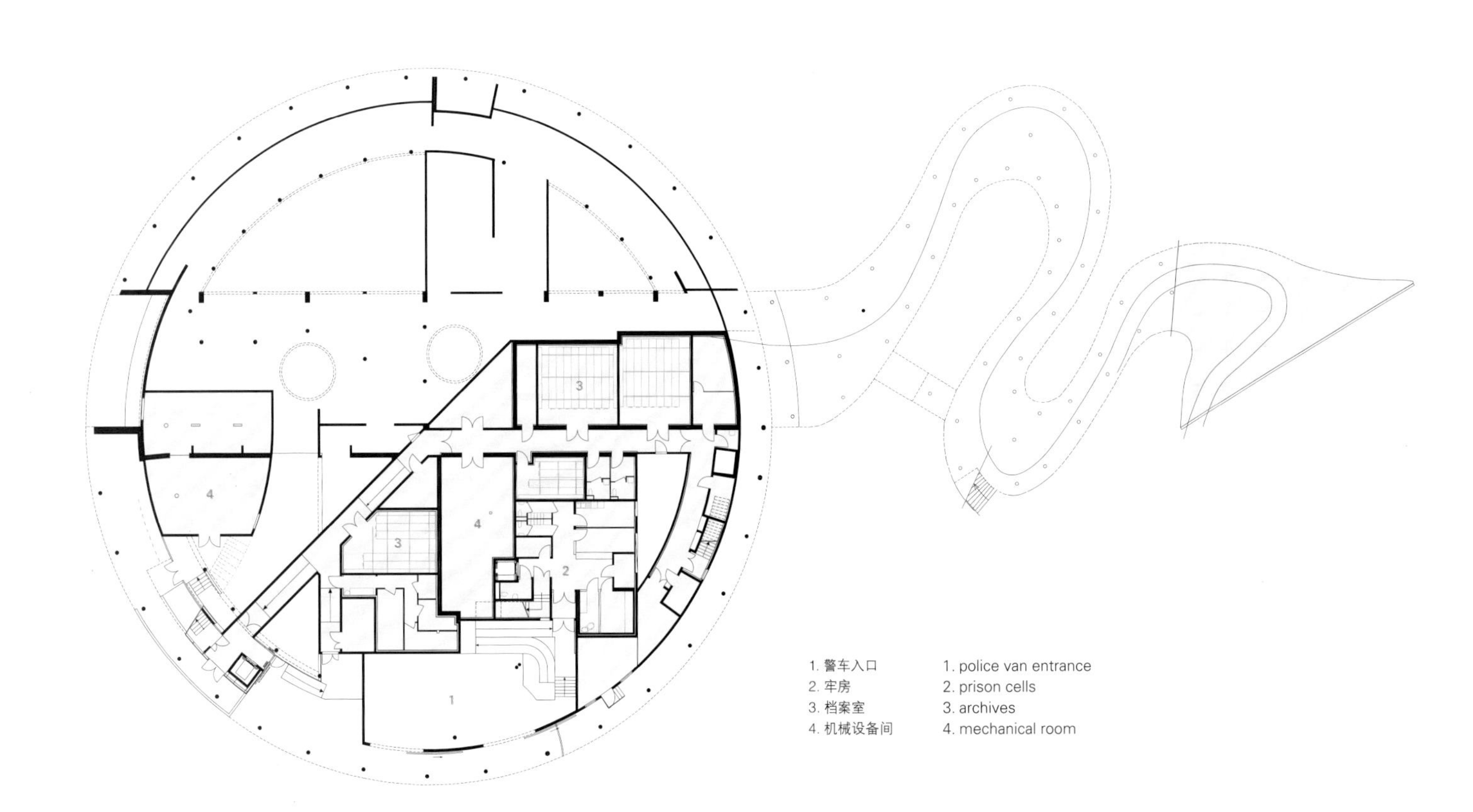

地下一层 first floor below ground

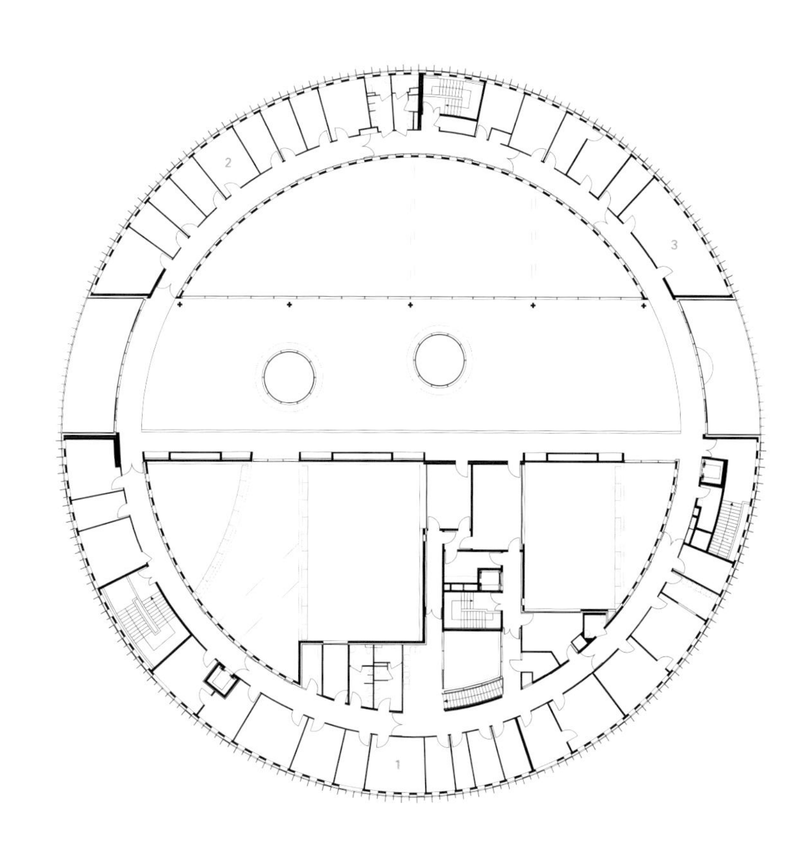

1. 检察官办公室
2. 地方法院
3. 劳资关系委员会

1. prosecutor's office
2. district court
3. labor relations board

二层 first floor

Salle d'Audience Montcalm

ACCUEIL

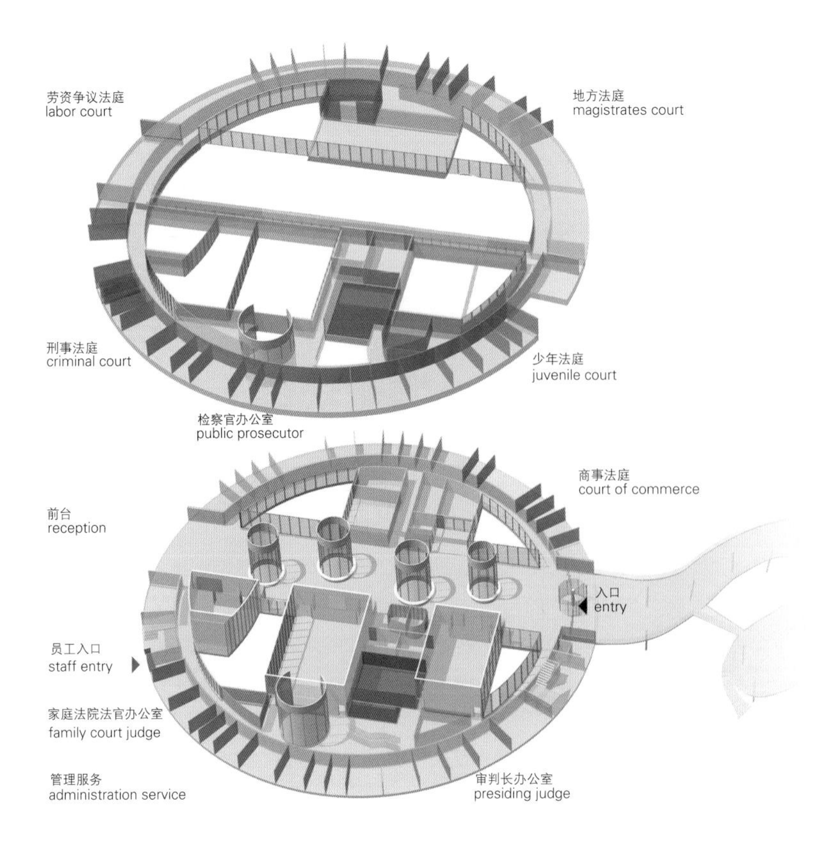
劳资争议法庭
labor court
地方法庭
magistrates court
刑事法庭
criminal court
少年法庭
juvenile court
检察官办公室
public prosecutor
商事法庭
court of commerce
前台
reception
入口
entry
员工入口
staff entry
家庭法院法官办公室
family court judge
管理服务
administration service
审判长办公室
presiding judge

SALLE D'AUDIENCE PYRÈNE

by the trees arranged in the plaza. They also give structure to and re-divide the linear crossing space of the "salle des pas perdues". The circular shape affords great functional efficiency; the two central staircases have been placed so as to minimize users' walking distance, allowing for centralized yet discrete access to the four jurisdictions. Courses could begin at any point, whether from the magistrates' access or the "salle des pas perdues". Finally, while the interior space can be easily modified, the architectural shape of the institution will remain.

The courtrooms are grouped on one side of the hall. The other side offers access to the public reception offices and the commercial court, which is independently accessible from the main entrance.

The courtrooms are located between the hall of steps and covers both levels of jurisdictions. The bay windows overlooking these planted outdoor spaces are treated with silkscreen frames so as to allow for a homogenous and glare-free indoor environment, and also to preserve privacy from the other spaces in the palace. The light wooden covering for the floor and false ceiling, as well as the furniture equipment, creates a clear and luminous atmosphere that lends an air of cohesion to the space.

The rest of the outdoor spaces have been designed with integration as the main priority. Integration with the surroundings is an extension of the project's driving goals of complete architectural coherence, restricted environmental footprint, minimized earthworks, and the revegetation of all rooftop areas.

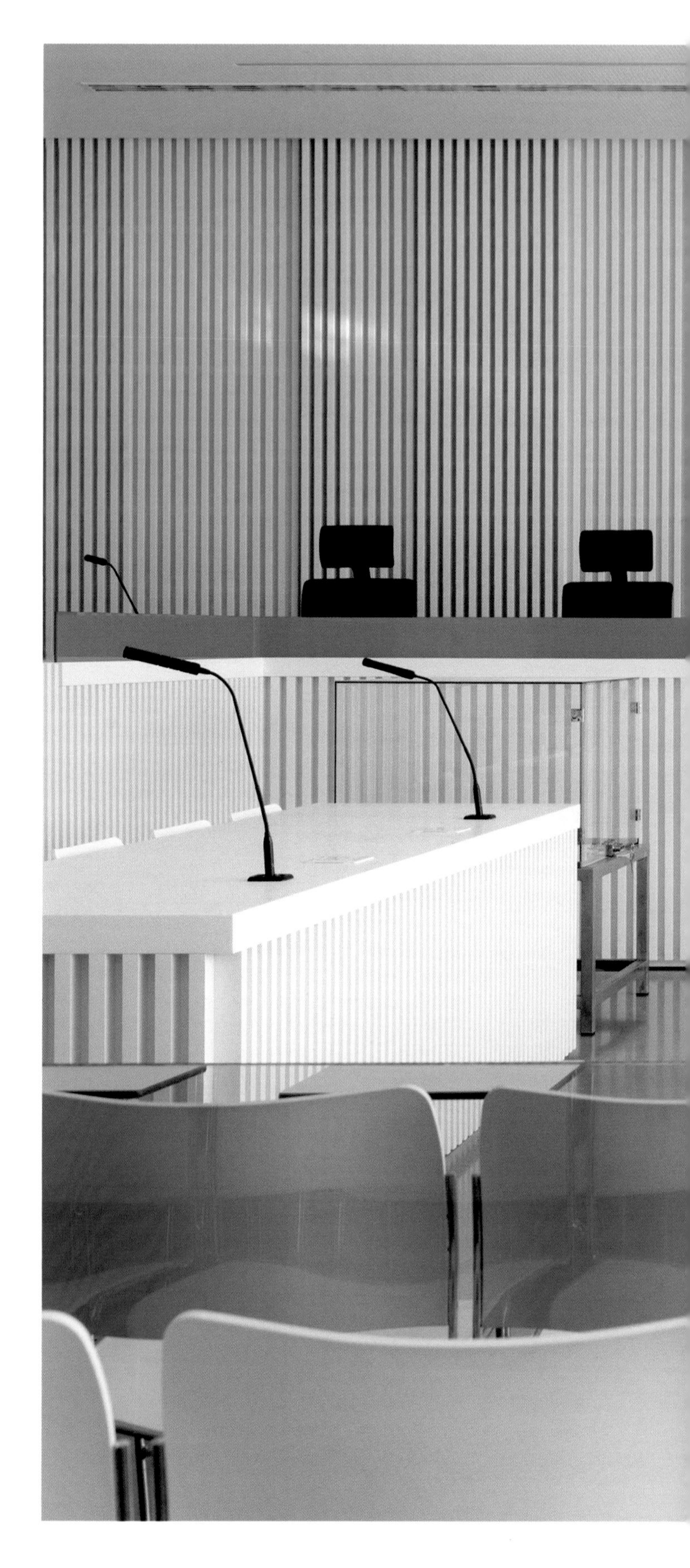

项目名称：Palais de Justice Foix / 地点：14 boulevard du Sud, 09000 Foix, France
建筑师：Philippe Gazeau / 客户：APIJ / DDT Ariège
任务：MOP Law (Law on Public project management), complete mission
经济学家、工程：BET TCE / Engineering - Sibat
承包商：Bourdarios - Vinci group; Hervé Thermique; Cofely Ineo
用途：court of first instance, magistrates court, court of commerce, labour court, offices, court hearing rooms, defendants areas / 建筑面积：4800m² SHON - gross area; 4320m² SDP - floor area; 2777m² SU - usable area
材料：concrete structure and steel framework; aluminium outside joinery; glass and anodised stainless steel sun screen / 能源效率：RT 2012 HQE
造价：€10,913 430 HT (excl VAT) / 设计竞赛时间：2011.6 / 竣工时间：2016
摄影师：©Gabrielle Voinot (courtesy of the architect) - p.173, p.174~175, p.177, p.182~183, p.184, p.186~187, p.189; ©Philippe Ruault (courtesy of the architect) - p.172, p.176, p.178~179, p.181, p.185

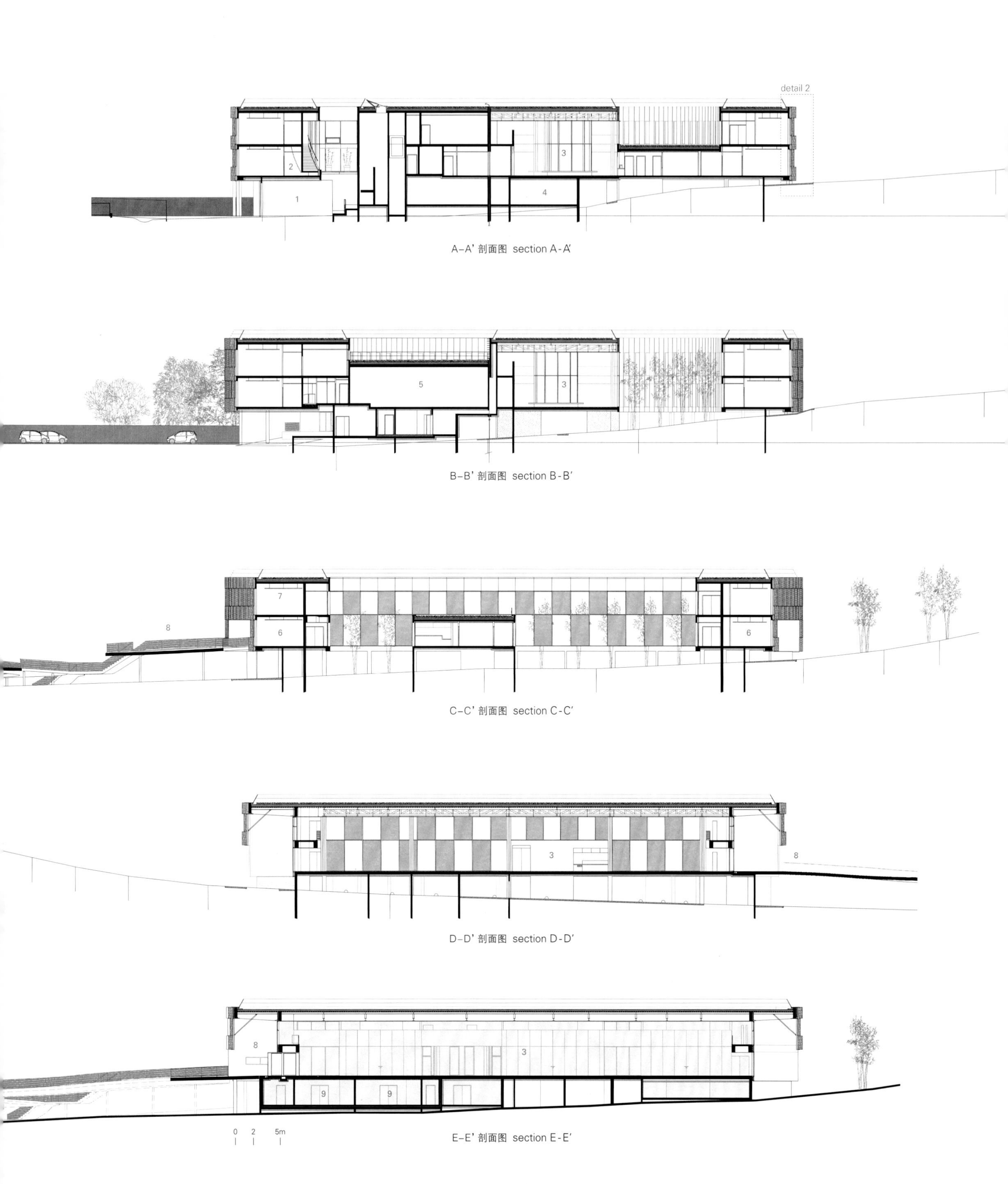

A-A' 剖面图 section A-A'

B-B' 剖面图 section B-B'

C-C' 剖面图 section C-C'

D-D' 剖面图 section D-D'

0 2 5m

E-E' 剖面图 section E-E'

1. 警车入口 2. 最高法官办公室 3. 等候厅 4. 机械设备间 5. 法庭
6. 商事法庭 7. 劳资关系委员会 8. 广场 9. 档案室
1. police van entrance 2. supreme court presidency 3. waiting hall 4. mechanical room
5. courtrooms 6. commercial court 7. labor relations board 8. plaza 9. archives

自由空间——第 16 届威尼斯建筑双年展

Freespace, the 16th Venice Architecture Biennial

Words by Tom van Malderen

©Jacopo Salvi

Giardini Central Pavilion

©Andrea Avezzù

©Jacopo Salvi

Arsenale

第16届国际建筑展开幕了，全球众多振奋人心的建筑项目皆来此参展。建筑展的策展人是伊冯·法雷尔和谢莉·麦克纳马拉，二人是位于都柏林的格拉夫顿建筑事务所的创始人，他们选择的参展项目多具有慷慨精神与人类的意识。格拉夫顿建筑事务所坚信，这些品质使建筑物具有维系与人之间有意义的连接的基本能力。法雷尔和麦克纳马拉将"自由空间"定为展览主题，并发表了宣言，为人们在这次规模庞大、多样化的展览中找到凝聚力提供了方法和指南。他们试图超越建筑物本身，将目光放在建筑内部，以"自由空间"为契机，引导人们将建筑视为一个充满机会的空间，无论是从字面意义上的还是隐含意义上，无论是情感方面的还是智力方面，都是如此。正如麦克纳马拉所说，"我很高兴地看到人们通过建造、调试或是感知来更加接近建筑，并有意愿成为建筑创作的一部分。""我们认为，人们对建筑需求的增长是促进其繁荣的关键。"

他们认为，建筑师或许没有能力与全球问题抗争，因此建筑设计重点应该放在许多微小的变化上，这些小变化会引起重大变化，并最终形成一种推动力。格拉夫顿邀请参展者和国家馆带着他们各自的

The 16th International Architecture Exhibition opened its doors with an inspiring and rich body of projects harvested across the globe. Curators Yvonne Farrell and Shelley McNamara, founders of Dublin-based studio Grafton Architects, assembled their selection around architectural examples of generosity and thoughtfulness. Grafton strongly believes that these qualities sustain the fundamental capacity of architecture to nurture and support meaningful contact between people and place. Farrell and McNamara established their manifesto and theme *FREESPACE* as a measure and a guide to finding cohesion within the diversity of an exhibition of this enormous scale. It is their attempt to move beyond architecture viewed as an object and instead, look at the in-between, imagining *FREESPACE* as an invitation to think about architecture as a space of opportunities, both literally and metaphorically, emotionally and intellectually. As McNamara states "I would love if people are brought closer to architecture and have a desire to be part of the making of architecture, whether through making, commissioning or awareness. We think that growing the desire for architecture is key to the flourishing of architecture."

Since architects may not have the power to battle global concerns, they argue, the focus should rest upon a lot of minor changes that can lead to something big and build up a momentum. Grafton invited participants and national pavilions to bring to Venice their own *FREESPACE*, to reveal

courtesy of La Biennale di Venezia

自由空间来到威尼斯，基于人物、地点、时间、历史展示出建筑的多样性、独特性和连续性。他们的宣言呼吁人们在这个充满活力的星球上保持住建筑文化与建筑实用性，并激励建筑师们对其工作的重要性和连续性持乐观态度。

军械库的制绳厂展厅和绿城花园中央展馆的主要展品令人振奋，这些项目充分利用了威尼斯的环境、氛围和地理位置所给予的慷慨馈赠，同时也印证了展会的宣言。"建筑并不是砌砖累瓦，"法雷尔说道，"而是为了找到一些元素来提高建筑的层次，将其从还不错的事物提升为有价值的事物。"制绳厂展厅宽敞恢宏的空间、红砖结构、不断变化的光线对比度以及中央展馆顶部明亮的灯饰，都是空间所给予的内在馈赠。格拉夫顿如今正在充分利用大自然的免费馈赠：光、空气、重力、自然资源和人造资源；他们充分利用已有资源和成功案例，重新发现其魔力，并提醒我们认识到它们的意义。

参观者一进入制绳厂展厅，就可以看到沿着中央长廊排列的展品，展馆建筑也被吸纳为展览的一部分。展品大多是三维模型，周围空间布置精巧，并配有步行浏览装置。在众多的参赛作品中都可以发现

the diversity, specificity and continuity in architecture based on people, place, time, history. Their manifesto is a call to sustain the culture and relevance of architecture on this dynamic planet, and inspire architects to be optimistic about the importance and continuity of their work.

The main exhibitions at the Arsenale's Corderie and the Giardini's Central Pavilion are inspiring testimonies to their manifesto, using the context and atmosphere of Venice and the locations as a generous gift. "Architecture is not building," said Farrell. "It's about finding the ingredients that lift something over the threshold, from being something that is okay to be something of value." The specific qualities of the Corderie with its heroic dimensions, its red brick structure and its changing light contrasts and the luminous quality of the zenithal light in the Central Pavilion are used as intrinsic gifts of space to share. Grafton are making the most of nature's free gifts of light, air, gravity, natural and man-made resources; they exploit the magic of what is already there, work with the things that are already successful and remind us of their meaningfulness.

Upon entering the Corderie, participant's displays are organised along a central promenade and absorb the building as part of the show. The displays are mostly three-dimensional, embedded in walkthrough installation and intricate spatial arrangements. An abundance of *FREESPACE* interpretations can be found amongst the many participants.

©Andrea Avezzù

Liquid Light/Flores & Prats

©Tom van Malderen

Close Encounter/16 younger practitioners

©Italo Rondinella

Dreams and Promises - Models of Atelier Peter Zumthor/Atelier Peter Zumthor

©Tom van Malderen

Field/Aireus Mateus

大量对"自由空间"的诠释。

里卡多·弗洛雷斯和伊娃·普拉茨在展厅中重现了他们在巴塞罗那的萨拉贝克特剧院项目的一部分，并整合了制绳厂展厅一个朝南的窗户。参观者们沉浸在他们的考古设计过程和现有建筑中挖掘出来的神奇空间中。该参展项目也是众多可以使参观者从建筑的历史积淀中获得馈赠的展览项目之一。在展厅的尽头，是一个由艾利斯·马特乌斯设计的梦幻装置，参观者受邀将头伸到一个UFO形状的黑钢多面体中。在里面，参观者会看到一个怪诞的干花花园，这不禁使人想到格拉夫顿对建筑慷慨的定义包括了对光、气味和触觉的使用，超越了建筑材料本身。

清除了经年累月的遮挡物和污垢之后，中央展馆获得了更好的通风和光线。威尼斯著名建筑师卡洛·斯卡帕设计的一扇被遗忘已久的窗户重新显露出来，将室内空间与室外的蓝天绿水再次衔接起来。在展厅内部，一系列迷人的模型和装置占据着巨大的中央空间。一层的展厅近距离展出了16位年轻从业者的作品，这些作品源自对艾琳·格雷和凯·奥托·菲斯克等过去著名建筑的反思。上层展厅是一个装有

Ricardo Flores and Eva Prats recreated a fragment of their Sala Beckett theater project in Barcelona and integrated a south-facing window of the Corderie. Behind it all, the visitors are immersed in their archaeological design process and the magical spaces they excavate out of existing structures. It is one of the many exhibited projects that extracts its gifts from the accumulated layers of history found onsite. Near the end of exhibition, a dreamy installation by Aires Mateus invites you to poke your head inside a UFO-like faceted black steel object. Inside, an eerie garden of dried flowers, reminds us that Grafton's definition of architectural generosity extends beyond the material, and includes the use of light, smell and touch.

The Central Pavilion was given additional air and light by clearing obstructions that had infiltrated over time. A long-forgotten window with interlocking circles by the great Venetian architect Carlo Scarpa was uncovered and connected the interior space once again with the water and sky outside. Inside, the large central space is taken over by a fascinating array of models and installations. Close Encounter at ground floor level shows sixteen younger practitioners with works that originate in a reflection on well-known buildings of the past by the likes of Eileen Gray and Kay Otto Fisker. The upper level is dedicated to a "workshop" of models with a captivating variety in terms of scale, appearance, and materiality, illustrating Peter Zumthor's journey from

澳洲馆 The Australia Pavilion

修复
修复是提倡建筑要促进或者积极参与建筑相关的自然环境、社会和文化方面的修复工作。这个理念和澳洲建筑师的工作环境尤为相关，因为他们处在世界上最为多样性、生态最为敏感的地区。他们在展馆里建了一片草地（植被修复）、装设了维持生命所必需的光源装置（日光），并将建筑通过录像展示出来（地面）。

Repair
Repair advocates architecture that catalyses or actively engages with the environmental, social and cultural repair of the context. It is particularly relevant to Australian architects who work in one of the most diverse and ecologically sensitive landscapes in the world. They create a grassland in the pavilion - Grasslands Repair, the life-sustaining light installation - Skylight, and present architecture through video - Ground.

©Tom van Malderen

北欧馆 The Nordic Countries Pavilion

©Italo Rondinella

另一种慷慨精神
如今人类已经进入"人类世"——我们已经征服了自然。在这个新时代，建筑应当能够定义一个完整的循环，从建筑最本质的组成部分到运行系统。"另一种慷慨精神"探讨自然与建筑环境之间的关系，以及建筑如何促进二者的共生共存。该展馆试图打造一种空间体验，以提高人们的环境意识。同时，北欧馆意在促进对话、讨论和批判，以帮助我们找到新的方法，以另一种慷慨精神来改造这个世界。慷慨精神不仅存在于人与人之间，同时也存在于人与自然之间。

Another Generosity
Humankind is now facing the Anthropocene – we have come to overpower nature. In this epoch, architecture should be considered as a tool for redefining the complete cycle of building, from its most essential components to its operating systems. Another Generosity explores the relationship between nature and the built environment, and how architecture can facilitate the symbiotic coexistence of both. It seeks to create a spatial experience which heightens our awareness of our surroundings. Moreover, it is an attempt to foster dialogue, debate, and criticism to help reveal new ways we can shape our world with another generosity. A generosity is not just between humans but between humans and nature.

模型的"车间"，这些模型在规模、外观和物质性方面都具有多样性，使人着迷。这些模型展现了彼得·卒姆托从思想和记忆到现实的旅程。作为一个致力于当代建筑实践的展览，中央展馆的档案馆里有许多令人耳目一新的建筑奇观。这使我们能够清晰地洞察到两位策展人的行为和思想，这正如雪莱·麦克纳马拉所说："我们并不认为时间是线性的，时间更像是古代和现代共存的一种螺旋体。"

今年参展的63个国家展馆以及在威尼斯全市并行举办的展览活动大部分都体现了"自由空间"的理念，鼓励游客重新审视他们的思维方式，并探讨看待世界的新方法。这些展览和活动对艺术展的主题进行了一系列不同的诠释，期间有很多问题被重复提及，值得我们讨论。格拉夫顿呼吁大家关注文化，例如，关注花园，以及将地球视为客户，这些理念在几个展馆中都有明显的体现。澳大利亚馆在展馆内外种植了一万多株植物，以提醒我们在占领土地时所面临的危险。北欧展馆充斥着巨大的膜结构，容纳着空气和水等基本元素。它强调了人类活动对地球行为的显著影响，并质疑了如今人类在自然界中的地位。阿根廷馆的一个神秘的立方体内有着延绵不绝的草原景观，这不禁使人思

thoughts and memories to actuality. For a show dedicated to contemporary architectural practice, there are a refreshing amount of architectural wonders from the archives to be found in the Central Pavilion. It provides an illuminating insight into what makes the curator duo tick, as Shelley McNamara states "We don't see time as a linear thing, but more like a spiral, where ancient and contemporary coexist." *FREESPACE* resonates within most of this year's 63 national pavilions and collateral events throughout the city, encouraging the visitors to review their ways of thinking and debate new ways of seeing the world. They offer a range of different interpretations on the theme and leave us with a number of reoccurring topics at the same time. Grafton's appeal to tend to culture, like tending to a garden and to see the Earth as Client, echoes visibly in several pavilions. The Australian pavilion installed ten thousand plants inside and outside of the pavilion to serve as a reminder of what is at stake when we occupy land. The Nordic pavilion is filled with giant membrane structures holding the basic elements air and water. It highlights the pronounced impact of human activity on the behaviour of our planet and questions our position in relation to nature today. An infinite pampas landscape is housed within an enigmatic cube at the Argentinian pavilion, making us wonder whether it is necessary to continue building and whether we succeed in improving the pre-existing through the constructions we build.

梵蒂冈馆(罗马教廷) The Vatican Pavilion (The Holy See)

Eduardo Souto de Moura

梵蒂冈教堂

罗马教廷馆，又称“梵蒂冈教堂馆”，以伍德兰教堂为模型精确打造，由建筑师冈纳·阿斯普里德于1920年在斯德哥尔摩公墓上建造。为了帮助参观者理解建筑师的设计，展馆的主入口处设有一个展览空间，展出了这座由阿斯普里德设计的教堂的模型及平面图。10位建筑设计师受邀围绕该主题进行建筑设计，他们将在威尼斯的圣·乔治·马焦雷帕拉迪奥岛上设计和建造10座教堂，这些教堂将共同构成梵蒂冈教堂馆。这项工程对建筑师们来说是一项不同寻常的挑战，因为该项目要求建筑既要孤立地存在，又能够融入一个完全抽象的自然环境。教堂建在小岛上的树林中，被泻湖所包围，10座教堂漫无目的地坐落其上，教堂周遭的环境与人生的漂泊起伏暗暗契合。

Vatican Chapels

The pavilion of the Holy See, named Vatican chapels, is based on a precise model, the "woodland chapel" built in 1920 by the architect Gunnar Asplund in the Cemetery of Stockholm. To help visitors understand this choice, an exhibit space has been set up at the entrance of the pavilion, displaying the drawings and model of Asplund's chapel. This theme has been proposed to the ten architects invited to build ten chapels, gathered in the wooded area in the island of San Giorgio Maggiore in Venice, to form the pavilion of the Holy See. The request addressed to the architects implies an unusual challenge, since the designers had been asked to come to terms with a building that will be isolated and inserted in an utterly abstract natural setting, characterised by its openness to the water of the lagoon. In the forest where the "Asplund pavilion" and the chapels have been located where there are no destinations, and the environment is simply a metaphor of the wandering of life.

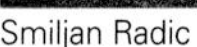

Smiljan Radic

Norman Foster

Ricardo Flores + Eva Prats

©Alessandra Chemollo

考我们是否有必要继续建造房屋，以及我们的建筑是否成功地改善了现有的生活。

梵蒂冈是威尼斯双年展的六个新参展国之一。该展馆占据了圣·乔治·马焦雷帕拉迪奥教堂后面的新景观林地，展示了建筑师设计的10个小教堂，意为“信仰、相遇、冥想和敬礼之地”。“自由空间”引发了人们对宗教及其空间影响的若干思考。建筑师在巴林馆中立起了一个充斥于整个房间的惊人的铝框架结构，并在一个大“荷电空隙”周围建造了走道。这一设计的中心思想是中东的星期五布道、公众集会以及备受质疑的公众自由。在绿城花园展厅，以色列馆分析研究了该国最重要的五个圣地，因为它们的宗教重要性，这些圣地经常成为苦战之地。一系列的建筑模型表明，不明确的领土主张创造出了一些最重要和最具挑战性的建筑地点。

政治联盟、政治组织和政治系统对建筑的影响在若干方面皆有所体现。比利时馆和英国馆都在审视欧盟的现状，并为其他国家提供“自由空间”，包括机会空间、民主空间、未经规划的自由空间和尚未构想的自由使用空间。英国馆的荒废暗示着英国的脱欧，但同时，该馆新建

One of the six new participants at the Venice Biennial is the Vatican. Occupying newly landscaped woodlands behind the Palladian church of San Giorgio Maggiore, it's showing 10 chapels by architects intended as a "place of orientation, encounter, meditation, and salutation". *FREESPACE* provoked several reflections on religion and its spatial impact. The Bahrain pavilion erected a startling aluminium framed structure that fills the room and creates a walkway around the edge of a large "charged void". The installation is centered around the notion of the Friday Sermon, the drawing together of people in public and the questionable public freedoms in the Middle-East. At the Giardini, Israel's pavilion examines five of the country's most important sacred sites, which have often become places of bitter struggle because of their religious importance. A series of architectural models show how uncertain territorial claims have created some of the most significant and challenging sites.

The impact of political alliances, organizations and systems returns in several contributions. Both the Belgian and British pavilion look at the state of the European Union, and offer a "free space" for others to take over; spaces for opportunity, democratic space, un-programmed and free for uses not yet conceived. The British pavilion, left abandoned, alludes to Brexit but equally reaches out to its neighbours with a newly created viewing platform on the roof that pokes through the trees like the survivor of a flood. Other highlights

捷克共和国馆 The Czech Republic Pavilion

©Tom van Malderen

联合国教科文组织
如今，捷克共和国的城镇Český Krumlov正面临着城市中心居民减少这一历史性问题。联合国教科文组织致力于解决这一问题，它为人们提供免费住宿，以及在市中心维持正常生活的工资。作为联合国教科文组织的驻地，该展馆不仅旨在引起人们对这一问题的关注，而且还给出了类似地区应对这一问题的具体方法。参观者在接待处可以浏览该展馆活动的列表。在展馆内部，参观者可以看到Český Krumlov小镇的现场监控画面，在那里，工作人员正试图为小镇的街道注入活力。

UNESCO
Today, the town of Český Krumlov, the Czech Republic, is undergoing the fate of depopulation of historic city centers. UNESCO is a company that strives to resolve this issue by providing free accommodations and a wage for living normal lives in the town center. As the site for UNESCO, the aim of the pavilion is not only to bring attention to the issue but to propose a concrete way how afflicted areas might handle it. At the reception desk, visitors are able to page through a catalogue containing a list of normal activities and within the pavilion, they are able to watch a live feed from Český Krumlov, where an attempt to breathe life into the town's streets is underway.

瑞士馆 The Switzerland Pavilion

金狮奖_瑞士 240：住宅导览
直到最近，没有家具的公寓内部照片才在建筑媒体上变得司空见惯。如果你对此感到讶异，那恰恰是因为长久以来，建筑师都是通过“抑制”建筑内部形象来预测其表面功能的。“瑞士240：住宅导览”通过建造夸张的与之相对应的“房屋之旅”来歌颂这种没有家具的建筑形象。在这次参观中，展馆的计划承诺对建筑规模、组织和功能进行控制，凸显了这一世界上最受欢迎的居住地的潜在可塑性。

Golden Lion _ Svizzera 240 - House Tour
Photographs of unfurnished flat interiors have only recently become commonplace in architectural media. If its increased prominence seems strange, it is precisely because the performance of interior surface has historically been predicated on the "suppression" of its image. Svizzera 240 celebrates the promise of the unfurnished image by constructing its theatrical counterpart, the house tour. On this tour, the plan's promises of control over scale, organisation, and function highlight the latent plasticity of the world's most popular habitat.

©Italo Rondinella

的屋顶观景台也向其邻国展馆延伸开去，观景台穿过树木，看起来像洪水中的幸存者一般。捷克馆也是该展会的亮点之一，该馆将展览空间改造成虚构的联合国教科文组织总部，其目的是使联合国教科文组织世界遗产遗址的废弃城镇中心回归到正常生活的轨道上来。旅游业带来了很多好处，但是当你所在的城市持续成为游客蜂拥而入的中心时，又会发生什么呢？韩国馆出于对国家前卫文化的恐惧，力图重新构建政治权力与想象力之间的隐秘故事，以及政治制度与乌托邦理想之间的矛盾。韩国馆的参展项目都是在20世纪60年代后期建造的，用来宣传韩国工业强国的地位。与此同时，这些项目也提出了一个问题，那就是韩国年轻艺术家和建筑师是如何通过对这些项目的四种当代诠释来使这些愿望与现代现实相适应的？

社会住房项目仍然是双年展上一个重要的议题。维多利亚和阿尔伯特博物馆将彼得和艾莉森·史密森设计并于近期拆毁的罗宾汉花园的一部分运往威尼斯，展示了建筑师对“天空中的街道”的构想，将他们对社区深思熟虑的理论与似乎失败的现实并列放置。重建后的废墟，与韩国艺术家徐道获新近受委托的项目一起，展示了由居民装修改

include the Czech pavilion, that transformed the space into the headquarters of a fictional UNES-CO company, whose aim is to return "normal life" to the deserted town centres of Unesco world heritage sites. Tourism brings many benefits, but what happens when your city is constantly the centre of a tourist invasion? With Spectres of the State Avant-garde, The Korean pavilion seeks to reconstruct a hidden narrative about between political power and imagination, and the contradiction between political system and utopian ideals. The projects featured were all built in the late 1960s to serve as propaganda for Korea as an industrial powerhouse, and raise questions about how these aspirations fit with the modern-day reality by means of four contemporary interpretations of those projects by young Korean artists and architects

Social housing projects remain an important topic of conversation at the Biennale. The Victoria and Albert Museum shipped a section of Peter and Alison Smithson's recently demolished Robin Hood Gardens to Venice, showing the architects' vision for "streets in the sky" and placing their thoughtful theories of community next to its supposedly failed reality. The reconstructed ruin is accompanied by a newly commissioned work by Korean artist Do Ho Suh, presenting a panoramic documentation of the estate's modular interiors as they have been adapted, decorated and furnished by residents. Latvia's pavilion explores the

比利时馆 The Belgian Pavilion

欧盟乌托邦
尽管布鲁塞尔欧洲区是欧盟的主要领土，是欧盟实体和象征意义上的锚地，但它对欧洲的集体认同几乎毫无贡献。出于形态、安全和经济方面的原因，它也不受公民倡议的影响。因此，比利时馆通过建造公共广场来解决欧盟地区民主和公民空间不足的问题。这本名为《欧洲之旅》的指南以空间线索和城市或建筑痕迹为基础，让人们一窥欧洲和布鲁塞尔可能成为的样子，以及建筑空间设计者可能会从这座超国家城市学到什么。

Eurotopie
Despite being the EU's principal territorial, physical and symbolic anchorage, the European Quarter in Brussels barely contributes to a collective European identity. For morphological, security and financial reasons, it is also impervious to citizen initiatives. The Belgian pavilion, therefore, addresses the deficiency of democratic and citizen spaces in the EU Quarter by offering a public agora. Based on spatial hints and urban or architectural traces, its guidebook, Voyage en Eurotopie, provides a glimpse of what Europe and Brussels may be and how space-makers may learn from the supranational city.

©Italo Rondinella

英国馆 The British Pavilion

©Italo Rondinella

特别提名_岛
"不要害怕；岛上充满了噪音、声音和甜美的空气，它们给人以快乐，而不是伤害。"
——莎士比亚《暴风雨》(3.2)
英国展馆在原建筑的屋顶上建造了一个新的公共空间。展馆本身是面向公众开放的，但却空空如也，人们只能从公众中心看到瓦片制成的屋顶顶部，暗示着其下方是一个沉没的世界。岛屿既可以是避难之地，也可以是流放之地，暗示着诸如气候变化、荒废遗弃、殖民主义、英国退欧、重建和避难所等主题。

Special mention_Island
"Be not afeard; the isle is full of noises, sounds and sweet airs, that give delight, and hurt not".
– William Shakespeare, The Tempest (3.2)
The pavilion constructed a new public space on the roof of the original building. The pavilion itself is open to the public but empty, with just the peak of its tiled roof visible in the center of the public, suggesting a sunken world beneath. Island can be a place of both refuge and exile, that hints themes like climate change, abandonment, colonialism, Brexit, reconstruction and sanctuary.

造的住宅内部模块的全景记录。拉脱维亚展馆探讨了日益复杂的社会中的共同居住问题，以及建筑在为人们提供共同居所方面的作用。一个奇怪的蒸汽模型阐释了能源消耗、地缘政治和集体决策之间的关系。在朱代卡岛上，小意大利（美国大城市的意大利移民区）尚未开放的展馆邀请双年展参观者参观了朱代卡岛社会住房综合楼内的一间公寓，该建筑由意大利建筑师吉诺·瓦莱于1986年设计完成，展示了一座出身卑微、预算低廉的建筑能为一个社区提供什么。展览结束后，临时展览画廊将空置5年，最终被重新改造成社会住房，重归威尼斯市民使用。

瑞士馆以其引人入胜的建筑设计斩获了"最佳国家馆金狮奖"。瑞士策展人带领参观者参观了其参展的住宅，通过夸张的建筑敏感性，你可以从一个局外人的角度看到典型的瑞士房产公寓的独特之处。他们渴望以新的方式来反思公寓的室内设计在塑造我们的生活和身份方面所起的作用。英国则获得了国家参与奖的特别提名，他们通过一个大胆的方案，用"虚无感"来创造"自由空间"，用作活动及非正式活动平台。爱德华多·索托·德·穆拉获得了颁发给最佳参与者的"金狮奖"。他展出的两张航拍图片准确揭露出建筑、时间和空间三者之间本

question of living together in an increasingly complex society, and architecture's role in providing for it. A strangely steaming model illustrates the relationship between energy consumption, geopolitics, and collective decision making. On the island of Giudecca, the Unfolding Pavilion from Little Italy invites the Biennale visitors to an apartment inside the Giudecca Social Housing Complex, completed by Italian architect Gino Valle in 1986. One of its aims is to show what a building of humble origin and budget might be able to offer to a neighbourhood. After the exhibition, the temporary gallery will be converted once again in a social housing unit, after five years of being unoccupied, and finally returned to the citizens of Venice as such.

The Golden Lion for Best National Participation was awarded to Switzerland for a compelling architectural installation. The Swiss curators take the visitors on a house tour, which offers an exaggerated architectural sensibility through which you see the peculiarities of typical Swiss real estate flats from the perspective of an outsider. They aspire to open up new ways of reflecting on the role the apartments' interior shell plays in shaping our lives and our identities. A special mention as National Participation was given to Great Britain for the Courageous proposal that uses emptiness to create a *FREESPACE* for events and informal appropriation. Eduardo Souto de Moura received the Golden Lion for the best participant for the precision of the pairing of two

韩国馆 The Korean Pavilion

©Italo Rondinella

国家先锋派的幻影

韩国馆探索了20世纪60年代建筑与国家之间的复杂关系，在那个年代国家为建筑师提供了展现其想象力唯一可行的平台。“国家先锋派”留下的遗产或残骸，为我们提供了另一种可行的方法，来建立一座充满政治色彩的城市建筑，而不是采取一个更规范或中立的态度，将自由空间与公共设施的概念混为一谈。

Spectres of the State Avant-garde

The Korean pavilion explores the complex relationship between architecture and the state in the 1960s when the state provided the only viable platform for showcasing imaginative prowess. The legacy of, or the wreckage from, the "State Avant-garde" may offer an intriguing possibility today for an alternate path toward a politically-charged urban form as opposed to the more normative, or neutralised concept of free space as a public amenity.

©Italo Rondinella

©Italo Rondinella

特别项目——应用艺术馆
Special project - Pavilion of Applied Arts

“罗宾汉花园：倒转的废墟”/维多利亚和阿尔伯特博物馆

应用艺术馆展示了罗宾汉花园的一部分，这是目前在建的几十个战后重建住宅项目其中的一个。批评者认为，重建项目无法缓解大城市中心地区的住房危机问题。而支持者则指出，政府通过重建项目为居民提供了经济适用房，改善了居民生活条件。因此，对于社会住房的未来，该展馆提出了一个亟待解决的问题。

"Robin Hood Gardens: A Ruin in Reverse" / Victoria and Albert Museum

Pavilion of Applied Arts exhibits a section of Robin Hood Gardens, one of the scores of post-war housing projects that are currently earmarked for redevelopment. Critics argue that regeneration is doing nothing to ease the housing crisis in the metropolitan centers. Defenders point to the provision of affordable housing and the improvement of living conditions. It asks urgent questions about the future of social housing.

©Tom van Malderen

©V&A Museum

Robin Hood Gardens by Alison and Peter Smithson, 1972

©Francesco Galli

质上的关系。自由空间的概念不言而喻，简单明了。“银狮奖”则颁给了一家有前途的年轻参赛单位——De Vylder Vinck Taillieu建筑师事务所。他们的项目表现出充分的自信，慢节拍的设计为未来打开了无数可能。由Paolo Baratta担任主席的威尼斯双年展理事会还根据策展人法雷尔和麦克纳马拉的推荐，授予建筑师、评论家、历史学家肯尼斯·弗兰普敦“金狮奖终身成就奖”。

第16届国际建筑展于5月26日向公众开放，一直持续到11月25日，展示了建筑复杂的空间性质，并将对好奇的参观者产生影响。展览规模庞大，时间紧迫，需要一段时间来消化和学习。“老人种下树的时候便知道他们没有机会在树下乘凉，这样社会才能发展壮大（前人栽树，后人乘凉）”，这是一句希腊谚语，今年的策展人在他们的宣言中也一再提到，在某种意义上，他们成功地在展览中种下了自己的树。

“自由空间”证实了所有建筑都对公共领域产生了积极或消极的影响。建筑设计至少是能够回报社会的。

aerial photographs, which reveals the essential relationship between architecture, time and place. *FREESPACE* appears without being announced, plain and simple. The Silver Lion for a promising young participant went to De Vylder Vinck Taillieu architects for a project that possessed a confidence thanks to which slowness and waiting allowed architecture to be open to future activation. The Board of Directors of La Biennale di Venezia, chaired by Paolo Baratta, also awarded a Golden Lion for Lifetime Achievement upon recommendation of Curators Yvonne Farrell and Shelley McNamara to the architect, critic and historian Kenneth Frampton.

The 16th International Architecture Exhibition opened its doors to the public on 26 May and continues until 25 November with a physical presence of scale and quality that communicates architecture's complex spatial nature and will certainly leave an impact on the inquisitive visitor. The exhibition is vast, with times overwhelming and can take a while to digest and learn from. "A society grows great when old men plant trees whose shade they know they shall never sit in" is a Greek proverb this year's curators have been referring to in their manifesto and, in a sense, they succeeded in planting their own tree with the exhibition. *FREESPACE* confirms that all architecture has an impact, positive or negative, on the public realm. The least architecture can do, is to give something back.

©Camille Gharbi

P126 **ANMA**

Ever since it was founded in 2001, ANMA's partners, Nicolas Michelin[left], Michel Delplace[right] and Cyril Trétout[middle], have been developing innovative studies and productions in the fields of architecture, urban planning, and landscaping. Forms a network of designers, researchers, and artists as well as spaces for working, exchanging ideas and experimenting. Based in Paris, Bordeaux and Beijing, they produce ultra-contextual urban projects marked by constructive rigour and an unflagging determination to use natural energies. Their approach does not depend on style or technique but mirrors an attitude that each project is is customised to reflect the site and its users. The agency is a partner in research programmes at the University of Montreal and the University of Shenyang in China and with Eurhonet, an international group of social housing operators.

P32 **EFFEKT**

Was established in 2007 by two partners, Tue Foged[right] and Sinus Lynge[left] in Copenhagen, Denmark, it is an architectural collaborative based in the fields of architecture, urbanism, and research. EFFEKT is the Danish word for impact. EFFEKT believes that architecture and urbanism are about creating a lasting positive impact on our surroundings and our planet. In recent years, they have been distinguished in both national and international architecture scene through several prestigious and award-winning projects, such as Livsrum Cancer Counselling center, GAME Streetmekka Esbjerg, ReGen Villages as well as some of Denmark's largest urban planning projects, Rosenhøj and Vinge.

Herbert Wright

Graduated in Physics and Astrophysics from London University, and has worked in software publishing, press analysis and arts administration. Writes about architecture, urbanism, and art for magazines and newspapers across Europe, and is contributing editor of UK architecture magazine Blueprint. Launched and curated Lisbon's first public architectural event 'Open House' in 2012. His first book 'London High (2006)' was about London high-rise, and later book projects include collaborations with Dutch architects Mecanoo and Expo 2015 Gold Prize designer Wolfgang Buttress. Delivers occasional talks, tours and discussions.

Tom van Malderen

His work ranges from buildings to objects, installations and exhibition design. His practice probes the intersections between art, design and architecture, and looks at the material gestures of everyday design and the construction of social space. After obtaining a Master in Architecture at Sint-Lucas Brussels, he worked with Atelier Lucien Kroll in Brussels and Architecture Project in London and Valletta. Since 2018, he works as an artist and exhibition designer based in Malta. He forms part of the contemporary art platform Fragmenta Malta and the Kinemastik international short film festival, and regularly contributes to international magazines. In Malta, his work is represented by the gallery Malta Contemporary Art.

Davide Pisu

Is an Italian architect, PhD candidate at the University of Cagliari and visiting PhD candidate at the University of Hertfordshire. He currently is leading research on the relations between norms and architectural form. His research interests are centred on architectural design and theory, with a focus on the architecture of information and knowledge. His architectural works include housing and public spaces, as well as an ongoing collaboration with C+C04 Studio.

P108 **Pesquera Ulargui Arquitectos**

Jesús Ulargui Agurruza was born in Logroño, Spain in 1965. Received his Master's degree in 1989 and Ph.D of Architecture in 2004 from Universidad Politécnica de Madrid (ETSAM). Has been a Lecturer at his alma mater since 1993.
Eduardo Pesquera González was born in Santander, Spain in 1964. Received his Master's degree in 1989 and Ph.D of Architecture in 2010 from ETSAM. Has been a Lecturer at his alma mater since 1995. Is a Grant Holder of the Academia Española de las Artes de Roma 1994-1995.

P172 **Philippe Gazeau**

Founded his architecture agency in 1984 in Paris, after graduating as registered architect (DPLG) at the Paris la Villette school of architecture. Created the urbanism agency FGP(u) - French Global Project of urbanism - of which he is the associate manager in 2004. Since 2012 he is member of the International Workshop of Greater Paris. Is currently teaching at the Paris Val-de-Seine, Ecole Nationale Supérieure d'Architecture. Has won the PAN 13 in 1984, Album of Young Architecture in 1985, special mention at the Mies van der Rohe Awards in 1994, the Grand Prix Award for Environment architecture of the PACA (Provence-Alpes-Côte d'Azur) region in 2003.

P138 **AYESA**

Founded in Seville by José Luis Manzanares Japón, AYESA Civil Engineering and Architecture is a Spanish company with more than 50 years of experience. Its scope of work covers all types of works, in the sectors of infrastructures, architecture and management, territory management, airports and ports. Their respect for the environment leads its designers not only to take care of the landscape but also to contribute to making it more beautiful by making form the key factor, based on the study of a particular function and a reasonable budget. Combines creativity and technical notions with wide experience in all phases that involve the conception and development of a project - from architectural competitions to technical assistance

P138 **Mecanoo Architecten**

Officially founded in Delft in 1984, is led by creative director/founding partner, Francine Houben[picture-above], design and research director/partner Dick van Gameren, technical director Friso van der Steen, and financial director Peter Haasbroek. The extensive collective experience results in designs that are realised with technical expertise and great attention to detail. Discovers unexpected solutions based on process, consultation, context, urban scale, and integrated sustainable design strategies, the practice creates culturally significant buildings with a human touch. Mecanoo's projects range from single houses to complete neighbourhoods, cities and polders, schools, theatres and libraries, hotels, museums, and even a chapel.

P154 **Renzo Piano Building Workshop**

RPBW was established in 1981 by Renzo Piano with offices in Genoa, Italy, and Paris, France. Has since expanded and now also operates from New York. Is led by 11 partners including Renzo Piano, founding architect and Pritzker Prize laureate.

Renzo Piano was born in Genoa, Italy in 1937 and graduated from Politecnico di Milano in 1964. Has collaborated with Richard Rogers from 1970 (Piano and Rogers), and with Peter Rice from 1977 (Atelier Piano & Rice). Has received numerous awards and recognitions including: the Royal Gold Medal, 1989; the Praemium Imperiale, 1995; the Pritzker Architecture Prize, 1998; the Gold Medal AIA, 2008. Was appointed senator for life by the Italian President Giorgio Napolitano in 2013 and received the Columbia University Honorary Degree in 2014.

P36

P36

P66 **Franklin Azzi Architecture**

Since the foundation in 2006, founding architect, Franklin Azzi, developed his skills in architecture, furniture, and urban design throughout France and abroad. He has always taken a constructive approach, working from the perspective of sustainability, efficiency, and clearness. Franklin Azzi Architecture[FAA] deploys a multidisciplinary team composed of some 40 collaborators including architects, interior designers, decorators, graphic designers, art historians, and also parametric design researchers. Thanks to being at the crossroads of these many disciplines, a process of collective reflection arises. This attention to transversality allows FAA to explore, among others: large-scale renovation programs for industrial buildings, but also new construction, or specific interior design for luxury stores. In addition, Franklin Azzi also collaborates with real estate developers, designing outstanding buildings treated with almost surgical precision.

P26 **KOKO Architects**

Was founded in 2000 by two partners, Andrus Kõresaar[right] and Raivo Kotov[left]. Comprised of Team KOKO (Architecture) and Team MOTOR (Exhibition Design), they specialized in Architecture, Interior Architecture, Landscape Architecture, Exhibition Design, Graphic Design as well as Multimedia Design. The work of the KOKO architects comes down to the context – history, environment, and people. We add as little as possible yet as much as necessary to amplify the uniqueness and distinctiveness of the site. KOKO was established in the wake of a successful international breakthrough - designing the Estonian pavilion at EXPO2000 in Hannover, Germany. In the last 18 years, they received a number of prestigious European awards.

©Alexandre Tabaste

P36 **Atelier van Berlo**

Was founded by Janne van Berlo[p.206-bottom, left] in Rotterdam, 2015. Is fascinated to create places that move people. Wants to initiate an experience that goes beyond functionality, technicality or aesthetics. Is also trying to make a space that surprises, inspires, warms and welcomes us and a place that is grounded in both its users and its surroundings.

P36 **Eugelink Architectuur**

Was founded in 2014, Eugelink Architectuur set up a team of experts for every assignment. The founder, Margriet Eugelink[p.206-bottom, right] is an all-round architect, who believes that a nice home, a pleasant well-designed neighbourhood or a fantastic workplace contributes to the well-being of people. She also devises building concepts like Healthy Ageing, Aktiv House, Cocreation, and Smart Cities. Under the motto of "let's uncover everyday aesthetics that mankind needs to feel good", they focus on Housing, Working, Caring, and second life: redevelopments.

图书在版编目(CIP)数据

司法建筑 : 汉英对照 / 意大利皮亚诺建筑事务所等编 ; 唐瑞译. — 大连 : 大连理工大学出版社, 2019.10

(建筑立场系列丛书)

ISBN 978-7-5685-2229-8

Ⅰ. ①司… Ⅱ. ①意… ②唐… Ⅲ. ①法院－建筑设计－汉、英 Ⅳ. ①TU243.4

中国版本图书馆CIP数据核字(2019)第221641号

出版发行：大连理工大学出版社

（地址：大连市软件园路80号　　邮编：116023）

印　　刷：上海锦良印刷厂有限公司

幅面尺寸：225mm×300mm

印　　张：12.75

出版时间：2019年10月第1版

印刷时间：2019年10月第1次印刷

出 版 人：金英伟

统　　筹：房　磊

责任编辑：张昕焱

封面设计：王志峰

责任校对：杨　丹

书　　号：978-7-5685-2229-8

定　　价：258.00元

发　行：0411-84708842

传　真：0411-84701466

E-mail：12282980@qq.com

URL：http://dutp.dlut.edu.cn

本书如有印装质量问题，请与我社发行部联系更换。

建筑立场系列丛书 01：
墙体设计
ISBN：978-7-5611-6353-5
定价：150.00 元

建筑立场系列丛书 02：
新公共空间与私人住宅
ISBN：978-7-5611-6354-2
定价：150.00 元

建筑立场系列丛书 10：
空间与场所之间
ISBN：978-7-5611-6650-5
定价：180.00 元

建筑立场系列丛书 11：
文化与公共建筑
ISBN：978-7-5611-6746-5
定价：160.00 元

建筑立场系列丛书 19：
建筑入景
ISBN：978-7-5611-7306-0
定价：228.00 元

建筑立场系列丛书 20：
新医疗建筑
ISBN：978-7-5611-7328-2
定价：228.00 元

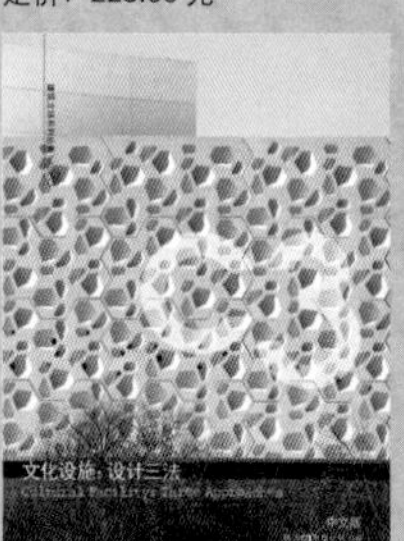
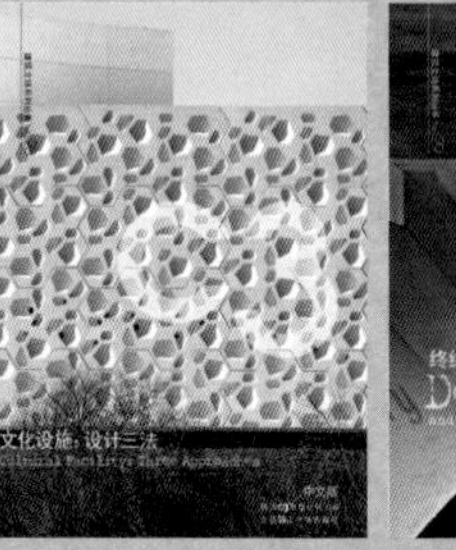

建筑立场系列丛书 28：
文化设施：设计三法
ISBN：978-7-5611-7893-5
定价：228.00 元

建筑立场系列丛书 29：
终结的建筑
ISBN：978-7-5611-8032-7
定价：228.00 元

建筑立场系列丛书 37：
记忆的住居
ISBN：978-7-5611-9027-2
定价：228.00 元

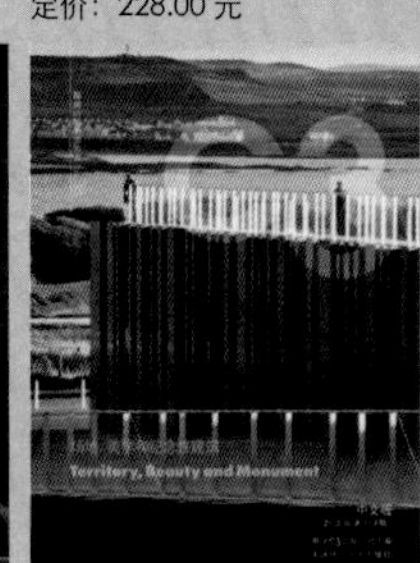

建筑立场系列丛书 38：
场地、美学和纪念性建筑
ISBN：978-7-5611-9095-1
定价：228.00 元

建筑立场系列丛书 46：
重塑建筑的地域性
ISBN：978-7-5611-9638-0
定价：228.00 元

建筑立场系列丛书 47：
传统与现代
ISBN：978-7-5611-9723-3
定价：228.00 元